Are We All Scientific Experts Now?

Dear First-Year Student:

On behalf of President Stuart Rabinowitz and the entire Hofstra community, I welcome you to Hofstra University. You are about to experience Hofstra PRIDE – an academic community of students, faculty and staff who will provide the foundation and support for your intellectual and personal growth over the next four years. Expect to be challenged, expect to be nurtured, and expect to be surprised and delighted by the new directions your life will take.

Your summer orientation program is designed to whet your appetite for your experiences this fall in a Hofstra classroom. To prepare for orientation, I ask that you read the class of 2020 common reading, *Are We All Scientific Experts Now?* The author Harry Collins explores the ways in which scientific claims can be unknowingly misrepresented in media and misunderstood by the public, sometimes with tragic consequences. He attempts to restore society's confidence in science by illustrating the need to take into account the varying levels of "expertise" possessed by the individuals or agencies who interpret and communicate data on national and international platforms.

To get started, visit the Common Reading website: hofstra.edu/ commonreading. There, you can view video clips of Hofstra faculty and students talking about the book, and you can share your own thoughts about it.

If you have questions about the Common Reading or about New Student Orientation, please call (516) 463-4874 or send an e-mail to orientation@hofstra.edu.

I look forward to meeting you this fall.

Sincerely,

Gail M. Simmons, Ph.D.
Provost and Senior Vice President for Academic Affairs

Are We All Scientific Experts Now?

HARRY COLLINS

polity

Copyright © Harry Collins 2014

The right of Harry Collins to be identified as Author of this Work has been asserted in accordance with the UK Copyright, Designs and Patents Act 1988.

First published in 2014 by Polity Press
Reprinted 2014 (twice), 2015 (twice), 2016

Polity Press
65 Bridge Street
Cambridge CB2 1UR, UK

Polity Press
350 Main Street
Malden, MA 02148, USA

All rights reserved. Except for the quotation of short passages for the purpose of criticism and review, no part of this publication may be reproduced, stored in a retrieval system, or transmitted, in any form or by any means, electronic, mechanical, photocopying, recording or otherwise, without the prior permission of the publisher.

ISBN-13: 978-0-7456-8203-7
ISBN-13: 978-0-7456-8204-4(pb)

A catalogue record for this book is available from the British Library.

Typeset in 11 on 15 pt Adobe Garamond
by Toppan Best-set Premedia Limited
Printed and bound in the United States by RR Donnelley

The publisher has used its best endeavours to ensure that the URLs for external websites referred to in this book are correct and active at the time of going to press. However, the publisher has no responsibility for the websites and can make no guarantee that a site will remain live or that the content is or will remain appropriate.

Figure 0.1 is reproduced from Wikimedia Commons (their history/Wikimedia Commons)

Every effort has been made to trace all copyright holders, but if any have been inadvertently overlooked the publisher will be pleased to include any necessary credits in any subsequent reprint or edition.

For further information on Polity, visit our website: www.politybooks.com

CONTENTS

Figures and Tables   *page* vi

Introduction: The Growing Crisis of Expertise   1

1 Academics and How the World Feels   17

2 Experts   49

3 Citizen Sceptics   80

4 Citizen Whistle-blowers   103

Conclusion: Are We All Experts Now?   115

Notes   133
Bibliography   138
Index   142

FIGURES AND TABLES

Figures

0.1 The Skylon   *page* 2
3.1 The target diagram   84

Tables

2.1 (Simplified) table of expertises   62
2.2 Question and answers concerning
     GW physics   70

The growing crisis of expertise

In 1951 my parents took me to the Festival of Britain. I stood under the 'Skylon' – a vertical, 300-foot, pointed, aluminium cigar, suspended 50 feet above the ground on steel cables. The Skylon seemed to float and it was a thrill for a kid to stand directly underneath, thinking that if scientists and engineers were not so clever the massive object would spear down through the top of my head.

A hundred years after the Great Exhibition, the Festival of Britain was meant to show Britons that they could recover from the war through the enterprise of the people, and the brave new world of science and engineering. Just a few years later, in 1956, the world's first commercial nuclear power station, Calder Hall, was connected to the national electricity grid. A year after that ZETA, the first ever fusion-power reactor, was completed. The radio, the newspapers and the newsreels (the Collins family couldn't afford a television) were saturated with these events. Today, the post-war cadences on old recordings still evoke that sense of a bright

Figure 0.1 The Skylon

future. Calder Hall had tamed the fearsome power of the atom bomb and used it to bring electricity into the home. ZETA was to tame the still more limitless power of the hydrogen bomb – the power of the Sun. In a speech in 1954, anticipating the development of fission and fusion power, Lewis Strauss, the Chairman of the United States Atomic Energy Commission, told the National Association of Science Writers: 'Our children will enjoy in their homes electrical energy too cheap to meter.' And along with Skylon there was nylon – shirts and blouses so much easier to wash – and jet airliners, penicillin, polio vaccine and new economic theories assuring the end of unemployment. These wonderful things were the products of *scientific* experts, remote, powerful and few in number.

But it all went wrong. Those nylon shirts were sweaty and turned yellow round the collar, the Comet jet airliners fell out of the sky, the new economics failed to prevent stagnation of the economy even in times of inflation. As for fission power, the more we learned about it the more we saw how it would leave a poisonous and financially crippling legacy to future generations – the vast cost and danger of disposing of radioactive waste. To someone of my generation, the modern idea that nuclear power is the saviour of the human race because it produces energy without warming the planet is a sick joke. What of fusion power? ZETA was an abject failure, embarrassingly closed down in 1961. In the 2010s, the hydrogen atom as a source of 'limitless energy' is as remote as ever and will have its own radioactive legacy if it ever works.

Since I was a kid it seems to have been one scientific or technological failure after another, played out before the eyes of every citizen. The future promised by the Festival of Britain soon lost its glitter. The cigar-shaped Skylon would be sold for scrap. Experts – what experts? Let us take a look at more recent events for which the fate of Skylon might be an icon.

In the UK in the 1990s, 'mad cow disease' or bovine spongiform encephalopathy (BSE), a disease located in the brain, spine and nerve tissue of cattle, took hold of the public imagination. It emerged eventually that it was caused by feeding dead cattle to live cattle. It may be that the slaughterhouses were also contaminated with remains of dead sheep suffering from a disease called 'scrapie' and this got into the meat and bone meal. Enforced cannibalism among herbivores is horrifying enough in itself and sinister when done in the name of improving the technical efficiency of farming. Hundreds of thousands of cattle were slaughtered, but still scientists insisted that there was no danger to the public from eating the meat of the infected animals with, famously, a British minister feeding a hamburger to his daughter on television. More than 150 Britons were to die a lingering death from the sinister 'Variant Creuzfeldt-Jakob disease' as a result of these mistakes. Science and technology both gave rise to the spread of the cattle disease and then visibly failed to understand it and its danger to humans, with commercial interests clearly swaying the publicly permitted pronouncements of the ministry scientists. Don't Jane Doe and the man on the Clapham omnibus feel they already know that feeding dead cattle to live ones can never be a good way of producing food for human consumption? How can this episode not encourage them to distrust or even campaign against the supposed benefits of science?

Only a little later, in 2002, we had the horrifying spectacle of thousands of cattle being burned on flaming funeral pyres throughout the UK as a result of a foot-and-mouth epidemic. Week after week scientists argued with each other about whether this was the right thing to do; should the cattle have been vaccinated instead? The public impression was of an incompetent government, informed by feuding scientists, squandering our farming heritage and failing to handle another risk to the food chain.

And all over the world, not least in the USA and other Western countries, people keep dying! That is to say, medical science continually fails to keep us alive and find cures for cancer or the thousands of other things we die from. An acquaintance of our family recently died, in a matter of days, from a small cut after it turned into necrotizing fasciitis; there is no treatment except cutting away the explosive invasion of infected flesh until there is no viable person left. And when we do recover from illness or surgery, mostly the body cures itself, sometimes better than others, because of the mysterious effects of mind on matter. We do not even know whether a new treatment works unless we eliminate the 'placebo effect' by using double-blind tests – the mysterious mind again. Newspapers are continually discussing the health benefits or dangers of almost anything that can be placed in the mouth or any other human orifice, and none of this builds into a pattern or is in any way consistent. When I was a kid most other kids were having their tonsils removed as a routine matter – no longer. Or is it no longer? Look at this from the *Washington Post* online on 25 April 2012, reporting from a medical conference:

> It turns out we're in the middle of . . . a tonsillectomy epidemic. . . . In 2006 alone, more than a half-million

children in the United States got their tonsils removed. The only problem is there's no evidence they work for most children.

Medical fashions swing from time to time and place to place – think of circumcision. Is it a science or is it a fashion influenced one way or another by religion or religious prejudice? The prejudice works sometimes in favour of circumcision and sometimes against. What can one do but choose one's preferred surgery oneself?

Or think of the science of economics. Econometric modelling is the most reliably unreliable predictor of national indicators we have. In the UK in the early 1990s, the government appointed a panel of 'Seven Wise Men' (econometricians) to advise them about the economy. Read from *The Independent* newspaper of 7 March 1993:

THE 'seven wise men' appointed by the Treasury to give Norman Lamont independent advice on the economy are publicly squabbling just a fortnight after publishing their first report. . . .

Professor Tim Congdon [has launched] a blistering attack . . . on the wisdom of his fellow panellists. One of the targets . . . described the assault as 'crazy and possibly libellous'.

So which of the seven turned out to be right? None of them! Commissioned and funded to build their most elaborate computer models for predicting next year's rate of inflation and next year's unemployment, they came up with completely different numbers. But surely the group as a whole was telling us something! Surely a crowd of economists do better than the individuals that make it up? No! The actual turnout was well outside the outermost prediction.[1] And now, at the time of writing (spring 2013), we are living in the aftermath of a financial crisis which, in spite of all the advice of economists employed by governments and their agencies, seemed to take everyone by surprise, not least the economists themselves.

Middle-aged citizens in the UK are also very fond of the famous statement by a much loved BBC weather forecaster, Michael Fish, who, at 9.30 p.m. on 15 October 1983, said in the course of his weather forecast:

> Earlier on today, apparently a woman rang the BBC and said she had heard that there was a hurricane on the way. Well, if you are watching, don't worry there isn't.[2]

A few hours later eighteen people were dead and 15 million trees had been uprooted as a result of 120 mph winds, the sort of thing that happens in the UK about once every hundred years. But that is short-term weather forecasting and the notoriety of the incident arises because, in general, short-term weather forecasting is pretty good. Long-term weather forecasting is usually accepted as hopeless, however. There simply is no viable science of long-term weather forecasting, though people keep trying.

The list above covers food science, veterinary science, medicine, economics and weather forecasting. Some of these are completely hopeless and some of them are hopeless in parts. Yet these are the sciences that affect our everyday lives in major ways. In the face of this, how is it that science ever achieved the status it once had?

The answer is that the science that has populated the imaginations of philosophers and other thinkers over the centuries is untypical in a statistical sense – quite unrepresentative of most of the science that goes on. It is heroic science. Newtonian science predicted the paths of known planets and even the existence of new planets from anomalies in the paths of the old ones. Einstein showed why the path of Mercury was not exactly as it should be even under Newtonian physics and a raft of previously unimagined physical effects were triumphantly confirmed. The quantum theory is said to be the most accurate theory ever, and it has quite counter-commonsensical consequences, thought absurd by Einstein, that turn out to be true – such as the instant 'communication' that happens under quantum entanglement. But all these triumphs in outer space and sub-atomic space happen where not much else is going on – both kinds of space are largely empty.[3] Understandably, when scientists want to advertise their craft they turn to the crown jewels of their profession, and when philosophers want a problem to mull over they also turn to startling successes, for there is nothing to explain about the scientific equivalent of a pair of torn underpants. Nevertheless, statistically speaking, when it comes to the problems faced by the citizen, it is nearly all drab garments at best, torn undergarments at worst. And this is the case because science does not cope very well with the crowded and complicated, and our world is crowded and complicated.

And things get worse. Science, we now know, is easily manipulated by cynical interests. The tobacco companies have paid scientists to argue that the link between smoking and lung cancer was less secure than the majority of scientists claimed.[4] It is only 'detective work' that can uncover the motives that lie behind the project because, on the surface, seamless science is being done: read the journals that contain the reports of the procured scientists and nothing seems to be wrong. It has also been shown that something similar has been happening in respect of climate change and the oil industry.[5] If our science is being made by puppet-masters pulling the financial strings of scientists, then aren't the opinions of Jane Doe and the man on the Clapham omnibus as good as anyone's?

Climategate

In 2009, Climategate hit the headlines across the world and in the US it led to an investigation in Congress. The University of East Anglia houses an important contributor to climate change science, known as the Climatic Research Unit. In late 2009, hackers motivated by climate change scepticism publicized email correspondence centred on the unit. Suddenly the public saw how scientists talk to each other when they are off the record, and it was horrifying: it did not look like science was supposed to look. George Monbiot, a high-profile environmentalist who writes for the newspapers, remarked: 'It's no use pretending that this isn't a major blow. The e-mails extracted by a hacker from the climatic research unit at the University of East Anglia could scarcely be more damaging.' In the US, on

2 December 2009, Jim Sensenbrenner, ranking Republican on the House Select Committee on Energy Independence and Global Warming, quoted from eight of the emails:[6]

From Kevin Trenberth:

The fact is that we can't account for the lack of warming at the moment and it is a travesty that we can't. The CERES data . . . shows there should be even more warming: but the data are surely wrong. Our observing system is inadequate.

From Phil Jones:

I've just completed Mike's Nature trick of adding in the real temps to each series for the last 20 years (i.e. from 1981 onwards) and from 1961 for Keith's to hide the decline.

From Andrew Manning:

I'm in the process of trying to persuade Siemens Corp . . . to donate me a little cash to do some CO_2 measurements here in the UK – looking promising, so the last thing I need is news articles calling into question (again) observed temperature increases – I thought we'd moved the debate beyond this, but seems that these sceptics are real die-hards!!

From Keith Briffa:

I tried hard to balance the needs of the science and the IPCC, which were not always the same. I worried that you might think I gave the impression of not supporting you well enough while trying to report on the issues and uncertainties.

From Phil Jones:

I'm getting hassled by a couple of people to release the CRU station temperature data. Don't any of you three tell anybody that the UK has a Freedom of Information Act!

From Michael Mann:

This was the danger of always criticizing the sceptics for not publishing in the 'peer-reviewed literature'. Obviously, they found a solution to that – take over a journal! So what do we do about this? I think we have to stop considering *Climate Research* as a legitimate peer-reviewed journal. Perhaps we should encourage our colleagues in the climate research community to no longer submit to, or cite papers in, this journal. We would also need to consider what we tell or request of our more reasonable colleagues who currently sit on the editorial board . . .

From Phil Jones:

If anything, I would like to see the climate change happen, so the science could be proved right, regardless of the consequences. This isn't being political, it is being selfish.

This looks like scientists fiddling the books, and more such damning emails poured out. The scientists, it emerged, were doing everything possible to keep their data hidden from sceptics; they were engaged in what looked very much like a political rather than a scientific battle. Even the scientific standards of honesty and openness, it seemed, were being sacrificed for the sake of victory in the case of scientists' opinions on the climate. The Climategate scandal went on and on, giving rise to endless public enquiries, and its meaning is still being discussed.

Thus has science fallen from grace and so has it become possible for celebrities and other campaigners drawn from the citizenry to appear on television, presenting their technological arguments in competition with those of scientists. And also, thus has the ordinary citizen come to feel that stepping out of the door will tell them just as much about the changing climate as a PhD or a professorial position. If we are asking whether we are all scientific experts now, it is Climategate that is the key to the trend: what happens inside science has become more visible to the citizen as television and the Internet have opened once hidden places to public gaze. Science's old image could never survive the examination, and when an idol falls it always invites a reaction.

In sum, since the middle of the twentieth century, science has been slipping from the high peak it once occupied and the citizen's relationship to science has changed. Citizens feel less intimidated when it comes to science and technology. Scientists have become less remote. Once they spoke from their mountain, offering revelations. Even television treated science as beyond us – the presenters would giggle over difficult technical terms, telling us they were way beyond the grasp of ordinary mortals, while the scientists were given licence to pronounce on anything. The television screen brought science into the living room, but only in the form of another holy icon. Now, as Climategate illustrates so well, the scientist in the living room is an ordinary mortal too. On the Internet, anyone can join in the conversation about, for example, the safety of vaccines. The experience of John and Jane Doe and their children is right up there with the Nobel Prize-winning research because the Nobel Prize-winning research has been done by people like you and me.

A new term is needed to describe this sense of empowerment – the sense that every citizen is part of the ball game of science and technology because there is no difference between us and the official ball players: let us call this feeling of empowerment 'default expertise'. Default expertise is the expertise – or, at least, right to judge – that ordinary citizens feel they possess because science and technology are so fallible. If experts do not really possess any special expertise – if 'the emperors have no clothes' – then, surely, 'we are all scientific experts now'. Possessing default expertise means being as good as an expert because there are no experts.

Whether ordinary citizens can really make scientific and technological decisions that are as good as, or better than, those made by scientists and technologists is a vital question. It has consequences for science, for our relationship with technology and for the future of our way of life. Maybe ordinary people's sense of what the weather is doing is a good guide to what is happening to the planet's climate over thousands of years, and maybe ordinary citizens can soundly gauge the weight of scientific evidence when it comes to vaccinating their children – and maybe not. We'll look in detail at the kinds of skill and abilities – the expertises – that specialists and ordinary citizens possess, and work out where the one can benefit the other and where it is only a dangerous illusion. We'll look at the conflicts among experts on this very matter of who is an expert.

ONE

Academics and How the World Feels

Why should the stream of scientific and technological happenings to which we have been exposed be made to add up to failure? We also see scientific and technological success on the television screen and the Internet. We watch scientific celebrities strutting their stuff: the late Carl Sagan staring at the stars from the deck of his spaceship, and ex-popstar and Professor of Physics Brian Cox evincing schoolboy wonder at the universe. We have seen Neil Armstrong stepping onto the Moon and we can watch satellite TV only because space rockets *do* work. Nowadays the journey to the airport is more dangerous than the plane ride. And the very Internet I use to get my anti-vaccination propaganda fix wouldn't be there without the scientists. Hasn't smallpox been eradicated and polio nearly so? Compare my teeth with my father's and grandfather's! So, in spite of ZETA, the Comet airliner crashes, *Columbia*, mad cows, tonsillectomies, mind over matter, medical fashions and our inability to cure necrotizing fasciitis, there is another way to add things up. The other way can be even more compelling in terms of the balance sheet of scientific benefit and loss. Why do we add it up one way rather than the other?

The way the public imagination adds things up depends on our overarching attitude to science and technology – it is something to do with the world-view, or spirit of the age – what we will call the 'zeitgeist'. The visible scientific failures and successes are the bricks; the zeitgeist is the mortar used to build them into a structure. Concentrate on the scientific successes and the wonder-struck heroes and you get one sort of building – a palace of science; concentrate on the failures and you get a hovel. In respect of science, anyone who lived through the second half of the twentieth century has seen the zeitgeist change along with so much else.

It all started in the 1960s when, along with the trilby hat and the sensible hemline, the long-established norms of every social hierarchy began to be challenged. At that moment science too began to lose its grip on the popular imagination. No one, aside from advertising agencies, press magnates and fascist dictators, knows how the zeitgeist works. I certainly do not. All we can do is look at indicators of what is going on. Academics are sensitive indicators: the arguments of academics tend to reflect what is going on in society as a whole. Academics also reinforce the popular world-view, feeding back what they take from society in popular books, top-end newspapers and television. I lived through the 1960s and 1970s as an academic, and watched myself and my colleagues drinking deep and refreshing draughts from the new way of living. 'The sixties' gave academics as well as everyone else a licence to think in all kinds of adventurous ways. I remember that sense of freedom and I remember that in 1972 it was not thought particularly strange that a paper published in the prestigious journal *Science* could argue that we needed a 'state-specific science' with its own unique findings that would apply only to those under the influence of LSD and other drugs.[7] I am not trying to set up a full-blown theory, something with hypotheses and tests; I am saying only that academic arguments can act as a litmus paper for the zeitgeist while feeding at least something back into it.

Whether or not it has been important, academics' reflection and reinforcement of the spirit of the age has been revealing. Since the 1960s, certain academic groups have been effectively trying to turn us all into default experts by showing that there is nothing special about science. For some this has been inadvertent, while for others it has been an explicit project. The academics in question come from the social sciences or the humanities and they make a living from reflecting on, researching and writing about the natural sciences. Since around the middle of the twentieth century there has been a boom in this kind of work – it is known as 'science studies'. I am one of its founders and long-term practitioners, making my first contributions in the early 1970s and still working in the field, so I can provide a participant's account.

Science studies began as a combination of a couple of fairly well-established disciplines, history of science and philosophy of science, along with my discipline – a relative newcomer – sociology of science. The history of all academic disciplines is immensely complex and it is impossible to say anything general about such things without glossing over details. Nevertheless, I am going to paint with a broad brush, dividing the history of science studies, or science and technology studies as it is often known, into three 'waves'. This is rather in the spirit that we might say that 'relativist physics superseded Newtonian physics' without worrying about the fact that Poincaré may have been thinking along Einstein's lines before Einstein, or that much of the physics we do today is still Newtonian. I know to my cost that many of those who make a living in science studies strongly object to finding themselves categorized as 'merely' part of a group, and they will hasten to point to the inadequacies of the simple picture I will paint.

But outsiders and those making their first acquaintance with a new field need to see the way the explanatory principles – 'the paradigm' – has changed over a generation or two. New students of the subject should be able to use the three-wave model as an introduction to the more detailed picture that they will encounter as scholars.

If, even as late as 1945, the Nazis had made nuclear bombs, they would have won the war and we would live in a very different world. And in the 1950s everyone knew it. In the 1950s, with radar, penicillin, nylon and all the rest, it was impossible to doubt the pre-eminence of science as a way of making knowledge. In this kind of atmosphere the job of history, philosophy and sociology of science was clear: explain how the scientific miracle worked. What was the secret of scientific knowledge and how could it best be nurtured? What the history of science did was research on scientific heroes while ignoring failures. The philosophy of science tried to explain how science reached certainties, with the typical problem concerning how observations articulated with theories. To use a standard example, a difficulty was that no amount of observations of white swans could prove that 'all swans are white' since the next one that turns up might not be white. A possible solution was that the observation of a single non-white swan could prove that 'not all swans are white' and this could be what counted as a scientific finding. After that it all became more complicated, but the idea was to crack the puzzle of science's success, not question it – that was the nature of the paradigm. The crucial point was that science could test its knowledge against a fixed world, and the only question was how that testing worked. The sociology of science in those days was looking at the way the scientific community had to be organized and what its values had to be for it to do its work successfully and to allow the testing to be carried on without bias or outside influence. It was said that democracy nurtured these values and that is why we, rather than the fascist dictatorships, had the science to win the war. These kinds of studies were driven on by their own momentum, becoming increasingly complicated and being subject to many esoteric and extremely clever arguments, but without ever challenging either the pre-eminence of science or the basic idea that, somehow, our scientific ideas were good because they were testable against reality. This is 'Wave 1' of science studies.

Then came the 1960s! In that decade it was suddenly possible to challenge any orthodoxy. A smallish book called *The Structure of Scientific Revolutions* was published in 1962; said to be the most widely read academic book of the twentieth century, its author was Thomas Kuhn. We now know that much of what was in that book was anticipated in another, then little-known book, published in German in 1935 and called, in English, *Genesis and Development of a Scientific Fact*. The author of that book was Ludwik Fleck, but to keep the story simple we will mostly stick with Kuhn.

One of the things that Kuhn pointed out was that science textbooks often contain little potted histories. For example, physics textbooks will explain relativity with a little story about the famous Michelson–Morley experiment, which was carried out in 1887 and produced a result showing that the speed of light was constant – a result which no one could understand until Einstein came up with relativity in 1905. This story is more of a myth than a real account of the unfolding of history. For instance, Michelson and Morley never completed their experiment to the point where it could be said to *prove* that the speed of light was constant – they gave up when they found they could not use the experiment to measure the speed of the Earth; arguments about whether the result of the Michelson–Morley experiment were correct went on right up to the 1930s. Furthermore, the experiment did not play much or any part in Einstein's thinking.[8] The potted histories at the start of textbooks are not, then, serious history, but little fairy stories. These fairy stories are, neverthe-less, a good way to teach the science. They do not do any harm to the scientists unless *policymakers* start to believe that science is really that simple. Actually, there is one more bit of minor harm: scientists who go on to do PhDs and other kinds of high-level research tend to get a nasty shock when they find that things aren't as straightforward as they had been taught and their job is really to organize a very untidy world – like trying to compress a balloon into a parcel tied with string. Of course, I am guilty of creating some simplified potted history of science studies right here, but I won't cause much harm either.

Let us say, then, that it was Kuhn's beautiful little book that had the most impact on the way historians, philosophers and sociologists of science began to think, as Wave 1 of science studies changed to what we will call Wave 2. One of Kuhn's theses, not found in Fleck's book, was especially appealing: this was the idea of the 'scientific revolution'. Kuhn said that certain new ways of thinking had a revolutionary effect on the way scientists interacted with the world. Einstein's idea of relativity is an example. Before relativity scientists thought of the world in a certain way: mass and energy were fixed and there was no limit on the speed at which things could travel. After relativity, mass and energy became interchangeable and, most remarkable of all, the speed of light became an absolute limit. Kuhn claimed this meant that for scientists, when the revolution took place, the way they acted – for example, how they did experiments and the conclusions they drew from them – changed too: for scientists, the very constituents of the world changed. (To complicate things, if one is looking for complications, this idea was anticipated by Peter Winch in his book *The Idea of a Social Science*, published in 1958, four years before Kuhn's book. It was inspired by the philosopher Wittgenstein, whose crucial ideas were published in 1953.)

Kuhn did not have any revolutionary ambitions in mind himself – he just thought he was putting forward an interesting slant on the history of science – but it often happens that those who take up an author's ideas use them in ways that he or she did not anticipate. The subtleties of the original argument can be lost because those who want to use a new idea need a simplified, punchy version. Academics often engage in a kind of journalism – they pick up the headline, not the detail, when they make use of another's work. The complications in the original get lost as we move away from the source, and things seem easy. A useful phrase to describe this and closely related processes is: 'distance lends enchantment'.

Authors sometimes find the crude use made of their ideas repugnant – certainly Kuhn did. Unfortunately for such an author, once the genie is out of the bottle it won't go back and, as we will see, the academic idea that there is nothing so special about science – the scientific emperors, if not completely naked, are in their underwear – is the great-great-grandchild of Kuhn's (or Fleck's, or Winch's, or Wittgenstein's, or . . .) genie.

Here is the point that emerges from Kuhn's idea of scientific revolutions: if the world changes in the course of a scientific revolution because of the way scientists think about it, then the world is no longer a fixed point. The world is no longer the anvil against which all theories can be hammered into shape. If *the world* changes when scientists think about it in a different way, then, not only *what counts as true* in science depends on where and when the scientists live, but what *is true* depends on where and when the scientists live: scientists who live in one place or time live in one world, while scientists who live in another place or time live in another world. The world can no longer settle the issue because it is one thing in one place and another thing in the other place. The anvil of reality has gone soft – like one of Salvador Dali's watches – and science is suddenly much more like religion, the arts, or all those other kinds of knowledge which up to the middle of the twentieth century had seemed inferior because, in contrast to science, they had no anvil on which their truths could be hammered out.

In the early part of the twentieth century there was an established field called 'sociology of knowledge', which dealt with how people came by their beliefs. It was easy to see that a baby born in, say, Southern Ireland was overwhelmingly likely to grow up a Catholic, believing that in the Mass the wine turned to blood and believing that they believed such a thing because the wine really did change. It was also likely that a baby born in Northern Ireland would grow up a Protestant, believing that in the Mass the wine merely symbolized blood and no transmutation took place, and believing that they believed it because no transmutation actually did take place. It never occurred to the grown-up Southern baby and the grown-up Northern baby that the ideas they were prepared to die for came to them merely as an accident of birth; the sociology of knowledge pointed out that this was mostly the case.

The sociology of knowledge is a dizzying subject: it leads one to start questioning all one's beliefs and to realize that pretty well all of them are a consequence of these kinds of contingencies rather than any real reasoning, however convincing things have seemed up to now. Up until the middle of the century, however, one could take a rest from this maelstrom of uncertainty on the calm island of the sciences and mathematics. Science and mathematics were immune from the sociology of knowledge because their island rested on the bedrock of reality. One sees, now, how it is that the idea of scientific revolutions causes so much trouble – the island is no longer fixed to bedrock but floats around in the ocean of knowledge, just like all those other kinds of beliefs that once seemed so comparatively insecure. In our simplified history, the softening of the anvil or the loosening of the foundations of science opened the way for the sociology of knowledge to invade the last redoubt. In the early 1970s, a new field called the 'sociology of *scientific* knowledge' began to grow. It became known by the acronym 'SSK' and was central to the growth of Wave 2 of science studies.

The 'principle of symmetry' was put forward in the early 1970s.[9] By this was meant that the social explanations of what are counted as true scientific facts and findings should be of the same sort as were used to explain false scientific facts and findings – just as with the Catholics and the Protestants. That, of course, is what the application of sociology of knowledge to science implies. Before then, an influential trend in philosophy took it that false facts and findings were to be explained sociologically, whereas the truth of true facts and findings was sufficient explanation in itself for why they were accepted, so there could be no sociology of true science. In short, we could have 'the sociology of error' but not the sociology of truth. SSK killed that idea, insofar as any philosophical idea is ever killed. From now on science was to be treated like religion or art: it would be a sociological disgrace to argue that, on the one hand, the Southern Irish Catholic believed the wine turned to blood because it really did – so no sociology was applicable to this true belief – whereas, on the other hand, the Northern Irish Protestant believed the wine only symbolized blood because they chanced to be born into a society where such things were mistakenly believed. In the sociology of religious knowledge 'asymmetrical' explanations of that kind would cause you to be kicked out of university in year one. But up to now those kinds of asymmetrical arguments had been perfectly acceptable if you were trying to explain scientific knowledge. From now on it would be as big a disgrace to provide asymmetrical explanations of scientific knowledge as of religious knowledge.

Empirical research inspired by the Kuhn idea (or the Winch, Fleck, Wittgenstein idea) was happening at the same time. For example, I was looking at the way scientists went about repeating experiments and confirming findings.[10] Replication of others' results, it seemed to me, and many others, was the key to the relationship between theories and the world. Neither Kuhn's talk of scientific revolutions nor the principle of symmetry would mean anything deep if scientists could continue to say 'we know what is true and what is false because anyone can replicate a valid experiment or observation whereas a false finding cannot be repeated and a false observation cannot be seen by others'. In that case the island of science would still be anchored to the seabed and the anvil of facts would be as solid as ever.

But experiments and observations are extremely difficult. When high-school students first look down a microscope at some pondwater, they see a mess; they do not observe the algae and the protozoa without being shown what to see. And when school students do experiments in the classroom they mostly fail until they are shown exactly how to get the right result. Exactly the same goes for professional scientists in the laboratory. It is easy to read the after-the-fact description of an experiment, but actually carrying out a similar experiment to the point of success is extraordinarily difficult – it is a matter of skilled manipulation and, usually, a lot of luck.

I had shown that scientists trying to build a new kind of laser – the TEA-laser – always failed unless they spent time in the company of a successful scientist; they had to pick up the knack of building it, a knack which neither party fully understood. This kind of invisible knack is known as 'tacit knowledge': it comprises the things we know but cannot say. If it was the case that TEA-laser builders – building a relatively robust device which produced a powerful beam of infrared radiation – always failed if they merely followed the published papers and circuit diagrams, how much more difficult would it be to use a novel and enormously complicated apparatus to confirm the existence of delicate and marginal effects?

As the 1960s turned into the 1970s, it was being claimed that just such a delicate and complicated apparatus had been used to detect the extraordinarily weak 'gravitational waves' – waves emitted, according to Einstein, when stars collide or explode. Others were trying to confirm or refute the claim and I went to talk to everyone involved. I found that when a group said they had looked and failed to see the waves, the discoverer responded that they had not built their apparatus carefully enough or said that they had run their experiment with too little care and dedication. Those failing to see the waves said the boot was on the other foot: it was the 'discoverer' who was making the mistakes as a result of sloppy experimentation and analysis. Since – as we have seen from the case of the TEA-laser – building scientific apparatus, like so much else, is a matter of tacit knowledge, it was very hard to settle the argument about who was right. In the case of the TEA-laser one could simply look to see if the infrared beam was present to know if one had succeeded in building the device because everyone agreed that is what a working TEA-laser should produce, but in the case of the gravitational-wave detector it was not clear what it should 'produce'. There was doubt about whether the waves should be strong enough to be seen at all, so the performance of the apparatus could not be used as a test of whether it was working properly; some scientists said a working gravitational-wave detector of the current sensitivity should detect gravitational waves, while some said it should not.[11] Thus, before one could use performance as a test of whether one had built a good piece of experimental apparatus and used it properly, one had to know what one should see with it, but to know what one should see with it one had to build a good piece of experimental apparatus and use it properly and so on, round and round; I called this 'the experimenter's regress'.

What happens in these cases, and they are frequent (remember cold-fusion), is that the experimenters split into two groups, one of which believes they can see the phenomenon and believes the other group are incompetent, while the second group thinks they have proved there is no phenomenon to be seen and the first group is incompetent. The outcome – what the majority of scientists come to accept – turns on all manner of 'non-scientific' considerations, such as who has the best reputation as an experimenter or what scientists' preconceptions are in respect of whether the phenomenon should be observed or not. Instead of the scientific facts being hammered out on the anvil of reality, they are formed through the normal social processes by which we come to prefer a political party or a style of art.[12]

This is how it is that scientific disputes often go on for so long. An extreme case is the small group of serious scientists, working in universities, who have been exploring the existence of paranormal phenomena – extra-sensory perception, psychokinesis and the like – for a century or more. They use scientific methods, the most up-to-date statistical analysis of their results and publish in journals that have the same appearance and standards as other scientific journals, and yet the large majority of scientists scorn them and their work. There are, as Kuhn might have said, should he have had any tolerance for parapsychological research, two scientific 'paradigms' at work here and two groups of scientists working alongside each other, seemingly living in different worlds, for decade after decade.

One might think that there is something especially strange about psychology and parapsychology but, as we have seen, similar things are to be found in physics. Nowadays, much new physics is first promulgated on an electronic preprint server known as 'arXiv' – the printed journals are too slow to keep an active physicist up with the research frontier. But, since its founding, arXiv has had endless trouble policing its boundaries. The administrators have had continual battles and even court cases, involving scientists who believe they have been unfairly prevented from promulgating their articles on the server. And these troublesome submitters, let it be stressed, are not unqualified – they are trained scientists with PhDs and many publications to their names, mostly working as professors in universities; they just have theories and findings that do not fit with what most of their colleagues believe. For this reason, the arXiv administrators have had to invent a special category into which these papers are pushed, with the insiders knowing that papers in 'general physics' need be taken less seriously than papers in the other categories. These outsider physicists are so sure they are being treated unfairly that they have founded their own journals. Of one article in such a journal, *Progress in Physics*, a mainstream physicist wrote to me that it was soundly put together:

It's professionally done. . . . The text is pretty good, the equations are mostly explained and the figures are clear. This man knows how to write a scientific paper.

But here is an extract from a 2009 editorial in that journal:

Declaration of Academic Freedom (Scientific Human Rights)

[From] Article 8: Freedom to publish scientific results
A deplorable censorship of scientific papers has now become the standard practice of the editorial boards of major journals and electronic archives, and their bands of alleged expert referees. The referees are for the most part protected by anonymity so that an author cannot verify their alleged expertise. Papers are now routinely rejected if the author disagrees with or contradicts preferred theory and the mainstream orthodoxy. Many papers are now rejected automatically by virtue of the appearance in the author list of a particular scientist who has not found favour with the editors, the referees, or other expert censors, without any regard whatsoever for the contents of the paper. There is a blacklisting of dissenting scientists and this list is communicated between participating editorial boards.

So science is not just one unified body of thought and findings – not even physics. Even in the hardest of sciences, one can find the equivalent of religious schisms. All this would have been puzzling under Wave 1 of science studies had it been looked at seriously. Instead there were, and still are, a plethora of articles, and even journals, that operate with the self-serving notion of 'crank' and resolve the problem with a snigger. This is poor stuff because the analyses mostly turn on the personal qualities of the scientists or some imagined methodological oversights, whereas at least a subset of the 'cranks' are indistinguishable from orthodox scientists in both their methods and their scientific integrity. Wave 2 explained how all this could happen and must happen; there is always enough room to interpret data in more than one way. That explains why the tobacco companies and the oil companies can pay scientists to produce results that go against the mainstream, and why the papers that come out of this false science can look just like papers coming out of true science; the only difference is the motivation. In sum, there is a great deal of 'interpretative flexibility' in even the hardest of sciences and the most careful of experiments, as becomes clear when the science is examined closely. The sociology of scientific knowledge, SSK, examined science much more closely than ever before. As a result, it turns out that we need to know motivations as much as we need to know results if we are to understand science!

The new approach to science which SSK initiated was very exciting. On the one hand, it led to many new and interesting findings which have enriched our understanding of science enormously. On the other hand, it fed into the long-standing rivalry between the sciences and the arts and humanities. In 1959, in his book *The Two Cultures*, C. P. Snow had complained that among the academic community that made up the arts and humanities, it was perfectly acceptable to display an ignorance of science and mathematics whereas it was nowhere acceptable to appear ignorant of matters humanistic. Nevertheless, it was the sciences that were pushing forward the frontiers of knowledge and these were difficult areas for the humanities people to penetrate critically. The perceived success of science in the 1950s threatened the old class-based hierarchy, therefore, but the new understanding of science resolved the uncomfortable paradox: science was not so different after all, so the arts and humanities had no need to feel inferior in the face of its success and nothing needed to change in respect of the status relationship between the two cultures.

The early SSK studies were hard and painstaking, it being thought necessary for the researchers to understand the science they were analysing. Then scholars started to describe the new kind of science studies in terms of literary criticism and semiotics – after all, what were scientific findings if not literary products, published papers that were favourably or unfavourably received? What did scientists do to establish their findings? They wrote and had their writing reviewed![13] This literary-based approach dominated science studies for a generation; now humanities and arts academics could criticize science even though they knew little of its substance. Distance, of course, lends enchantment so the critiques could become sharp and punchy – as one slogan put it, following Clausewitz, science is 'a continuation of politics by other means'.

We see that for increasingly wider groups of academics, by the middle of the 1980s, there was less and less special about science. For those influenced by these academics – and the influence became stronger after the arts and humanities discovered the literary critique of science – the bar had fallen to the ground and we could all be scientific experts. I remember one meeting where an artist explained to me that the problems of gravitational-wave physics would be solved if only the research teams were expanded to include the arts.

Other less esoteric streams fed, and continue to feed, into this ocean of ordinariness. First, it was argued that interests from the wider society fed into scientific conclusions. That such a thing could happen was made clear by SSK. If the decision about which set of experiments comprised the competent replications turned on the assessment of the scientists as well as the findings – and the interpretative flexibility of data as exemplified by the experimenter's regress made it inevitable – then the choice of scientist and, therefore, the choice of scientific finding, could well be influenced by what type of person the scientist was or which choice produced a favoured set of results. For example, feminist analysts argued that science was dominated by males and revealed that this showed through even into the actual substance of, say, reproductive technologies. It was not just that scientific conclusions *could be* influenced by interests outside the strictly scientific; it was that the outside world *had* fed into science's conclusions and, in this case, male bias at all levels, including findings, gave rise to one type of reproductive science rather than another.

Second, it was shown that when the products of science were put to use, scientists tended to work from an over-simplified version of the world and failed to take account of expertise based on experience.[14] Thus, farm workers were told that the herbicide 245T was safe so long as the appropriate safety precautions were followed, but the workers were the only ones with experience of using the substance and only they could say that it was impossible not to violate the precautions in use and that the herbicide was not safe in the way the scientists said it was. The farm workers did say exactly this but, without professional qualifications, it was hard to get their voices heard.[15] Cumbrian sheep farmers suffered from the same problem in being heard after the fallout from the Chernobyl disaster contaminated their land. The UK government scientists offered advice designed to minimize the effect of radioactivity on the sheep that was quite impractical, whereas the sheep farmers understood much better what needed to be done.[16]

Unfortunately, the authors of such studies said they had shown that the farmers possessed 'lay expertise' and the idea spread that laypersons' technical expertise had been shown to be as good as that of scientists. In some academic circles it became seen as a sign of an unacceptable elitism to suggest that scientists knew things that laypersons did not. The mistake made by the authors and their followers was that the farmers and farm workers were not laypersons but members of an elite – the elite of experienced farmers and farm workers. They were not members of a rich and powerful elite, nor even an elite in virtue of formal training and qualifications, but, as much as any group of scientists, they were an elite in terms of experience and expertise. It is no surprise, then, that these 'experience-based experts' had much to offer when it came to the practical application of scientific advice. In these cases the scientists acted unacceptably in excluding the farm workers – they acted like an elite that had power because it has elitist qualifications, but this power was used to exclude a rival elite from the debate. In spite of the injustice, there were still no 'lay experts' involved: members of the general public were nowhere to be seen, except in the distorted interpretations of the research. To call the rival elite 'lay experts' was to risk pandering to a sentiment for democracy that was in no way justified by the case studies themselves. Discovering small groups of specialist experts where no one had looked before was very valuable work but, among academics, the misleading description of what had been discovered nurtured the 'default expertise' position.[17]

Another important case study that fed into this way of academic thinking was of the revolt by San Francisco gays against double-blind testing of anti-retroviral drugs to alleviate AIDS.[18] The most dramatic action of this community was to share out the placebos and the active drugs that were distributed for the double-blind tests. The experimental subjects argued that the active drugs might save lives and they preferred to share the drugs and sabotage the test rather than act in a way that would allow only half of them to benefit from the potentially life-saving treatment, even if this was said by the medical profession to be the only way to test it.

The AIDS activists also took part in street demonstrations and other kinds of political activity to try to change the drug-testing policy, and eventually they succeeded. The medical profession did change their minds and found new ways to test the drugs that did not leave part of the at-risk population untreated. The AIDS activists contributed to the development of the new testing regime.

The author of this AIDS study has argued fiercely that the case demonstrates the way that laypersons can contribute to science. His argument seems less than secure, however, because the excellent case study reveals that even as the activists were mobilizing political support they were also learning the science of testing. At the end of the day a proportion of them had learned enough of the science to be able to argue fluently with the medical doctors and researchers, and it was undoubtedly the new skills they had acquired that enabled them to contribute to the development of the new kinds of test – while it also led them to understand and admit to the necessity of double-blind testing in certain circumstances. In this case the political activity of laypersons led to a reconsideration of the science, but the redevelopment of the science itself was done by experts – the original medical researchers and the new body of experts who had once been laypersons but were now metamorphosed into specialist experts through their continual interaction with qualified specialists. Therefore, once more, the author's interpretation of his own work seems to miss the mark but the case study, backed up by what he says about it, has certainly led many to believe it demonstrates something close to the idea that we are all scientific experts. Unsurprisingly, there is interpretative flexibility in social science as well as natural science!

Going back to the influence of the zeitgeist on academics, since the 1960s two other factors seem to have fed back into the way academics do their analysis before they feed back out again into the general public consciousness. One influence is an apparent clash between expertise and American political thinking where the conception of democracy is narrow enough to imply that experts are beyond the control of the people or beyond '"government by discussion" in which discussion is largely intelligible'.[19]

[T]he presumption in democratic societies is that all decisions should be as far as possible public; it is the exceptions that require justification. The American Freedom of Information Act, a crowning achievement of law in the Vietnam era, reflects just this sensibility. Second, public engagement is needed in order to test and contest the framing of the issues that expertise are asked to resolve. Without such critical supervision, experts have often found themselves offering irrelevant advice on wrong or misguided questions. Third, as we have seen, expertise is constituted within institutions and powerful institutions can perpetuate unjust and unfounded ways of looking at the world unless they are continually put before the gaze of laypersons who will declare when the emperor has no clothes.[20]

Thus, under this view, where the knowledge of experts is treated as esoteric, decisions made by experts cannot be questioned through normal democratic processes and this is abhorrent. To everyone's relief, however, the circle can be squared if the public are wise enough to say when the bar is on the ground; this makes the idea popular, especially among academics. Unfortunately, it is not clear from what base of expertise the public is meant to make these judgements. Is it that we are all default experts? Or is it that we are all specialist experts?

When analysts who favour the view that the public has expertise examine professional expertise, the tendency is to say that an expert is simply someone who is called an expert. In other words, when it comes to professional expertise, it is not what you know but what others think you know that counts. This is a 'relational theory' of expertise; one's expertise is made out of one's relationships with others. The only persons to whom this does not apply are members of the ordinary citizenry who do have genuine expertise in virtue of their experience and wisdom. Effectively, it is not just that the bar is on the ground but, bizarrely, that the roles of expert and citizen have been reversed.

One of the most powerful and prominent contributors to recent debates is also clear about the duty of academics to serve the zeitgeist, rather than question it. She insists upon it, not in the spirit of the way we might say, 'we were all affected by the sixties', but as an imperative – public consciousness should guide academic thinking:

> The worldwide movement in legislation and public policy these days is toward, not away from, wider participation . . . In general, Western states have accepted the notion that democratic publics are adult enough to determine how intensely and in what manner they

wish to engage with decision-making, subject only to the constraints of time and other resources.[21]

Perhaps this kind of reverence for the public accounts for the determination of analysts to read case studies in a way that shows how ordinary people can resolve the supposed 'paradox' of democracy and expertise if they are determined enough, while also making it clear that people do not need any special qualities, or to be in any way less ordinary, to contribute to technical decisions. What these analysts miss is that there are other models of democracy, which require only that decision-makers are accountable to the people through regular elections while being tasked with making decisions between elections in ways which are not immediately open to discussion. 'More democracy' is a siren call for academics – who can disagree? What is better for democracy than more of it? But without some carefully considered limits, the demand for more democracy is worth no more than a call for more motherhood and more apple pie.

The environmental movement is another influence on the way the world is generally understood to have changed since the 1950s. Science and technology's exploitation of natural resources, once viewed simply as a producer of wealth, has come to seem evil and dangerous. And, of course, in many ways it is dangerous even if not evil. In many places we have poisoned the Earth locally and we now know we have damaged the Earth as a whole, by, for example, creating a hole in the ozone layer and by warming the climate. The justified worry about the effects of science and technology feeds back into an almost religious prohibition of certain kinds of action, which influences and is influenced by academics. In Britain, for example, the North Sea oil rig *Brent Spar* came to the end of its life in 1995. The best option seemed that of sinking it to the seabed, but environmental groups opposed this move and were supported by academics. Shell bowed to the pressure and cut up the rig on land. Later analysis, accepted by all parties, showed that disposal at sea would have been the most environmentally friendly option; the supposed specialist expertise of the pressure groups had been misplaced.[22] This, however, did not stop public sentiment against 'polluting the ocean' continuing to drive at least some of the academic argument. The tendency continues in the case of academically supported public distrust of the use of Genetically Modified Organisms (GMOs) for food. We can add to this the general tendency of a good proportion of academics in the social sciences to favour the underdog and to be suspicious of the powerful. That is often how sides are chosen and, given interpretative flexibility, it is a strong influence on how evidence is put together.

Interim summary

I have tried to use academic debates as a kind of litmus paper for changes in the public understanding of what science means in our lives – where science stands in the zeitgeist of 'the West'. I have argued that scientific experts and the knowledge they produce have become much less valued since the 1960s. The academic field known as 'science studies' showed us how not to be trapped by science's experimental and observational methods into believing that scientific knowledge was superior to every other kind of knowledge, and these arguments were taken up by the arts and humanities who may have influenced the public in turn. To the extent that these ideas have filtered out into the wider world, their consequence will have been to reflect back into society the changes that began in the 1960s, making it still easier for us to believe either that we are all default experts, or that we all have specialist expertise.

TWO

Experts

So far we have discussed the notion of the 'default expert' and mentioned the closely related 'relational theory of expertise' – the idea that an expert is simply someone who is called an expert. In a sense, a default expert is not really an expert at all because their sense of being an expert comes only from the fact that there are no real experts. The supposed experts have been defined as experts because of the way they fit into social life – the relational theory – but the default expert sees through it. Some of us have come to see ourselves as default experts, either because professional experts have revealed their lack of expertise through constant failure, interpreted in a way encouraged by the changing zeitgeist, or because, in the case of a few, they have taken the lead from certain groups of academics who also reflect and reinforce the zeitgeist.

The relational theory of expertise is ingenious and insightful but it cannot be all that we need. This is because there are practical problems that it does not solve. The practical problems do not prove the relational theory wrong any more than bumping into a wall proves that life is not a dream, but the relational theory simply does not address the practical problems. For instance, it provides no guidance on how to choose between competing experts: if all experts are there only because people call them experts, how can anyone choose one over another? The relational theory also fails to capture the day-to-day experience of life. Learn or, more pertinently, fail to learn, to ride a bike, play the piano, read and write, or even speak a language, and the felt experience is of not having an expertise as opposed to having one. Maybe that experience is an illusion, but for 'all practical purposes' the experience has to be taken into account or we would not know what to do next. Should we take our bikes out on the road? Should we audition to be concert pianists? Should we buy some books for the beach? Do I need a dictionary for my foreign trip? Default expertise and the relational theory leave these questions unanswered. Therefore our experience of real expertises – *substantive expertise*s – has also to be thought about when we analyse the citizen's role in technological decision-making and the way we live our lives and make our choices. Who, then, are the substantive experts?

Models of expertise

The prevailing view among psychologists and philosophers is that an expert is someone who has devoted about 10,000 hours self-consciously developing their expertise. Think of a concert violinist, or a racing driver, or, since this is our topic, a scientist. They all fit the model. So we could stop the argument now: we are not all scientific experts, or even experts at anything, because very few of us have put in the 10,000 hours.

The trouble is that insisting that every expertise must take 10,000 hours or more of self-conscious effort gives rise to a problem: under this model, what counts as an expertise will vary from place to place. I am a native English speaker and, because the 10K model is widespread, my English is not generally recognized as an expertise in England. But, if I go to a foreign country where English is not the native language, my English *is* recognized as an expertise! If I go to live in France, I can command a salary and respect for teaching English. On the other hand, if you are born and brought up in France, it does require self-conscious attention to learn English. Thus, if all expertises are defined according to something like the 10K model, then expertises are always appearing and disappearing depending on where a person is – England or France, for example. They are also always appearing and disappearing depending on the circumstances. When word processors were first invented, pioneers in their use, such as myself, were counted as experts. In the very early 1980s, in the University of Bath's School of Social Sciences, where I worked, I was the second user of the department's first computer, a Tandy TRS80 with 8.5 inch floppy disks! The machine was bought for the economists to do their calculations, whereas I used it for word-processing – I was a pioneer. Then more computers were bought (with smaller floppy disks), other members of the faculty followed my example, and my achievements came to seem less remarkable. In the very early 1990s, as a Head of that School of Social Sciences, I had to work very hard to convince the administrative staff that it was not only the faculty who were clever enough to use computers to compose their texts; they were too. Just a year or so later, and even the most junior people in the administrative office were using word-processing computers and teaching the faculty how to do it. But the substance of the skills had not changed through all these transformations so why should the meaning of the expertise?

Ubiquitous expertise

What is missing from the psychologists' model is the notion of *ubiquitous expertises* – expertises that we have acquired without putting in any self-conscious effort. Ubiquitous expertises include speaking your native language, knowing table manners, how often to wash in this society or that, and how close to other people one should come when passing them on a path. Everyone acquires ubiquitous expertise as a result of growing up in a particular society and these things differ from society to society and from time to time.

Because ubiquitous expertises are acquired just through growing up, and because almost everyone in society acquires them, including the laziest and least intelligent, the existence of these expertises has been overlooked. But some of the expertises, such as the informal rules of one's native language, or knowing how close to come to others in a variety of circumstances, are very hard to express and learn – they are, in another sense, among the very hardest of expertises to learn. That a skill everyone can acquire could be difficult came as a huge surprise to the pioneers who tried to teach natural language to computers. It turned out that the kind of mathematics that even geniuses struggle with can be executed by machines, whereas we are still failing to make computers that can understand ordinary speech! Ubiquitous expertises have the following in common with default expertise: in respect of them we are all experts. But there is nothing of substance to default expertise whereas acquiring ubiquitous expertises is a very substantive accomplishment which we just happen to learn without putting in any self-conscious effort, or even noticing that it is happening. Notice that some expertises change their status over historical time from specialist to ubiquitous, or near ubiquitous. Word-processing is an example and car-driving is an even better one.

'Nobelskigrad'

Would it be possible, in principle, for us all to become experts now in a specialist science were we to put enough effort into it, just as the Bath University office staff became 'experts' in word-processing? Could it be that people would stop thinking of science as a special kind of accomplishment? Could science fade into the background of ubiquitous expertises? Or is it that scientists are special people so that science will always be the prerogative of an elite?

The answer is that, as in every other activity, there are certain people in science who are very clever.

Scientists like to tell stories of famous colleagues who would see their way instantly through a complex problem, while the lesser mortals around them could spend months struggling with it. Here are some descriptions of one of the most famous of these geniuses, John von Neumann:

> Von Neumann's ability to instantaneously perform complex operations in his head stunned other mathematicians. Eugene Wigner wrote that, seeing von Neumann's mind at work, 'one had the impression of a perfect instrument whose gears were machined to mesh accurately to a thousandth of an inch'. Paul Halmos states that 'von Neumann's speed was awe-inspiring'. Israel Halperin said: 'Keeping up with him was . . . impossible. The feeling was you were on a tricycle chasing a racing car.' Edward Teller wrote that von Neumann effortlessly outdid anybody he ever met and said 'I never could keep up with him'. Lothar Wolfgang Nordheim described von Neumann as the 'fastest mind I ever met', and Jacob Bronowski wrote 'He was the cleverest man I ever knew, without exception. He was a genius.'[23]

But this is not the essence of science – after all, no one said such things of a still greater scientist, Albert Einstein. Furthermore, it is said of another unchallenged genius of twentieth-century physics, Niels Bohr, that he was a poor mathematician:

> Niels Bohr's brother Harald, a great mathematician, once said that Neils could get along without mathematics because of his great intuition. Heisenberg declared that 'mathematical clarity in itself had no virtue for Bohr'. . . . There is ample evidence that Bohr's mastery of mathematics was very limited . . . all the mathematics of Bohr's great 1918 paper on atomic theory was done by Bohr's assistant at the time . . . 'Bohr had no idea how to do the mathematical part. Even though the physical ideas, so to say, he had.'[24]

In any case machines are taking over the calculation part of science, leaving the scientist to be the person with the foresight or insight, and there is no clear relationship between foresight and insight and exceptionally high IQ of the calculative sort. In looking for von-Neumann-type brilliance as we try to answer the question of whether we are all scientific experts now, we are barking up the wrong tree. There is no question that we are *not* all *brilliant* scientists now. Fortunately, to know enough to be able to make good scientific judgements is not the same as being brilliant.

Having disposed of the geniuses, let us ask the question again: could science be a 'ubiquitous expertise'? Let us invent a piece of science fiction or, to put it more grandly, let us try a thought experiment. Imagine one of those science cities built in the days of the Soviet Union, but more extreme. In 'Nobelskigrad' all the families are composed of scientists – mothers and fathers both – and scientists do all the menial work in their time off, such as collecting garbage, putting out fires and cleaning the drains. This community of scientists talks only science, day and night. Children born into the community hear only talk of science from the moment they are born – indeed, what they heard in the womb before they were born were faint echoes of science talk. The toys are test tubes and microscopes and the nursery walls are decorated with equations, mobiles representing DNA splitting and combining, and models of molecules rather than nursery rhymes. The school curriculum, from kindergarten onwards, is science. We can imagine that in this society, science would be a ubiquitous expertise and only the most intellectually challenged would fail to become a fluent mathematician, physicist, biologist, or whatever.

For the purposes of this book, then, let us agree that we have disposed of the idea that science is the preserve of the few in virtue of its intellectual difficulty. Science, in at least some of its forms, is no harder than being a handyman or handywoman. Furthermore, throughout the centuries, relatively ordinary people have contributed to science: gathering and observing specimens for natural history, watching the skies on behalf of astronomers and, nowadays, using their computers to analyse data in the searches for extraterrestrial intelligence ('SETI') and gravitational waves ('Einstein at Home').[25]

To conclude, there is nothing special about expertise that requires that it take 10,000 hours of *self-conscious* effort, and there is nothing special about scientists that makes science necessarily the preserve of an elite. Nevertheless, in our society there are no Nobelskigrads and this makes science into something other than a ubiquitous expertise: it is, under the circumstances in which we live, a *specialist expertise*. In our society, one does not become 'a scientist' without practice, and a lot of practice.

Specialist expertise and meta-expertise

In the last section the idea of ubiquitous expertise was introduced. At the end, the notion of specialist expertise was also mentioned. Specialist expertise is what is possessed by a doctor, a concert violinist, a carpenter, a physicist, a mathematician, a truck driver, an engineer and so on. 'Specialists' are mostly what people, including professional psychologists, are thinking of when they talk of experts.

There is also a third kind of substantive expertise, which is very important for the relationship between experts and public. This is 'meta-expertise'. Meta-expertise is used to judge and choose between other experts; in principle, this kind of expertise can be good enough to guide one through the decisions one has to make in the contemporary technological world. Faced with a conflict of technical opinion, if one has a sure method of choosing between the experts then one has a sure method of choosing the best option.

A table of expertises

Default expertise is a funny category because there is nothing to be learned except that the so-called specialists don't truly know anything. Lurking among the substantive expertises there is, however, a theory of acquisition. Our ubiquitous expertises are learned without self-conscious effort – they just drift into us. At least some of our meta-expertises are going to come to us this way too – we learn to discriminate between trustworthy and untrustworthy people as we grow up in our society and we would be 'clueless' in a society strange to us. A theory of acquisition of *specialist* expertise has been encountered too – with the TEA-laser. The ability to build a TEA-laser (if you were not the genius who invented it in the first place) depends (or, at least, in the early days depended) on social interaction with scientists who had already built a working laser. Only through social interaction could scientists pick up on the unspoken and, perhaps, unspeakable, knack of making the device work. It may be that nowadays more of the art has been expressed in a formal way, or the device itself has changed, but when one is in the arena of novelty or uncertainty – when one is dealing with things that are not as well-developed as your fridge or your car – practical success in science turns on tacit knowledge which is not so far from the tacit knowledge of the handyman or woman, or the tacit knowledge acquired when ubiquitous expertises are learned. The theory is that, while something can be learned from instruction books and other kinds of literature, the heart of an expertise is acquired by picking up the tacit knowledge. This can be made to happen only by hanging around with those who already have it, either the specialists or, in the case of ubiquitous expertises, the ordinary people. Where science is concerned, you always have to hang around with the specialists unless you live in Nobelskigrad.

Notice that what has been expressed here is a 'social theory' of the acquisition of expertise: it comes through mixing with other experts – it comes from social contact. The relational theory of expertise is also a social theory – becoming an expert is a matter of your social relations with those who assign the label 'expert'. But these are two very different social theories: the relational theory is about labelling; the social acquisition of tacit knowledge theory is about how to acquire a substantive expertise from other substantive experts. Both theories are to do with social interaction but that is where the resemblance ends.

We now have three kinds of substantive expertise – ubiquitous, specialist and meta-expertise – and a theory of how they are acquired. Table 2.1 is constructed around the idea that an expert is someone who shares the tacit knowledge of a specialist group. As can be seen, its three rows correspond to the three kinds of substantive expertise. The advantage of a table is that it invites subdivisions and keeps everything organized. As regards the question of whether we are all scientific experts, the two most important things are the heavy division between primary source knowledge and interactional expertise and the emboldened entry 'local

Table 2.1 (Simplified) table of expertises[26]

1. UBIQUITOUS EXPERTISES				
2. SPECIALIST EXPERTISES	*UBIQUITOUS TACIT KNOWLEDGE*			*SPECIALIST TACIT KNOWLEDGE*	
	Beer-mat knowledge	Popular understanding	Primary source knowledge	Interactional expertise	Contributory expertise
3. META-EXPERTISES	*EXTERNAL (Transmuted expertises)*		*INTERNAL (Non-transmuted expertises)*		
	Ubiquitous discrimination	**Local discrimination**	Technical connoisseurship	Downward discrimination	Referred expertise

discrimination'. These will be explained as we work through the categories again in a more detailed way.

Ubiquitous expertises

Row 1 of table 2.1 is the starting point. Ubiquitous expertises are, as has been seen, the expertises we all acquire from growing up in our society. If we had grown up in Nobelskigrad the ability to understand science would be among our ubiquitous expertises but, generally, to be a good scientist is to be a specialist expert.

Notice that whatever kind of expert you are, you start with the ubiquitous expertises that pertain to your society and you remain a ubiquitous expert even when you add other expertises on top. If we knew how to 'weigh' the total amount of expertise we each possess, we would find that our ubiquitous expertises hugely outweigh our other kinds of expertises. Thus, if you are a top physicist such as John von Neumann or Albert Einstein, you will have learned to speak a natural language, learned how often to wash and how close to walk to others. You will probably be literate and, nowadays, know how to use the Internet. All these are remarkable accomplishments and most seem impossible to reproduce unless you are a human.

Specialist expertises

What is it to have a specialist expertise and to what extent is it possible for a good proportion of us to acquire specialist expertise? Well, most of us are specialist experts in the substance of our work, whether it be carpentry, bus-driving or high-energy physics. We have already encountered specialist sheep farmers and specialist herbicide sprayers. Some of those normally thought of as members of the lay public are specialist medical experts. For example, those with chronic diseases have knowledge about the treatment of those diseases that compares with or even exceeds that of their doctors. Chronic disease sufferers can have a lifetime of experience of their illness and its treatments; in respect of each particular illness they comprise a knowledgeable elite. Once more, though drawn from the lay population (aren't we all), they are not *lay experts*: by the time they have suffered and treated their illnesses for a few years, they are just *experts* – experience-based experts.

Experience-based experts such as these should be recognized as belonging with the rightmost group of line 2 of table 2.1 – they belong to the contributory experts. A contributory expert is someone who makes a contribution to an area of expertise and is, generally, what people think of when they hear the word 'expert'.

How does one become a contributory expert? By working with other contributory experts and picking up their skills and techniques – their tacit knowledge of how to do things. One becomes a contributory expert by being an apprentice – an apprentice scientist, an apprentice farmer, an apprentice chronic disease patient, learning from other patients and from doctors how one's disease is to be managed. That is why the grey heading over the right-hand side of line 2 of the table indicates that to join this group requires 'specialist tacit knowledge'. We'll return shortly to the second category under this heading – interactional expertise.

There are three categories on the left-hand side of line 2 in table 2.1 where it is indicated that only *ubiquitous* tacit knowledge (that is, ubiquitous expertise) is needed for their acquisition. That is because all three of these categories depend only on *reading*, not on mixing with the specialist experts – they do not depend on any kind of apprenticeship. Reading, of course, does rest on a lot of ubiquitous tacit knowledge: one must know one's native language, one must be literate and one must know what to read and where to find it. To understand how much knowledge is involved one need only imagine a native from the Amazon rainforests being set down in the UK or the USA and told to research some technical topic by reading from the library or the net. The crucial thing about these three categories is that the ubiquitous tacit knowledge acquired through regular socialization, along with hard work, is enough. No specialist tacit knowledge is needed; it may be that these categories ought to be referred to as levels of specialist *information* rather than expertises.

The leftmost category of these three, 'beer-mat knowledge', requires least tacit understanding and less ubiquitous expertise than the others, though even reading and understanding a few lines of text is a remarkable accomplishment. The term, 'beer-mat knowledge', is based on the few lines of writing that might be found on the back of a beer mat or coaster. Thus, a decade or so back I found the following printed on a beer mat made by the Babycham company:

> A hologram is like a 3-dimensional photograph – one you can look right into. In an ordinary snapshot, the picture you see is of an object viewed from one position by a camera in normal light. The difference with a hologram is that the object has been photographed in laser light, split to go all around the object. The result – a truly 3-dimensional picture!

So now I am an 'expert' on holograms. One cannot use beer-mat knowledge for much – perhaps it is useful in the game of Trivial Pursuit but one cannot use it to help one make holograms, or make decisions about whether holography is a useful technique and should be supported, or decide whether holograms are good or bad for one's health or the environment.

The next category in the second row is 'popular understanding'; it is the knowledge that can be acquired from popular science books, or television shows. Once more, no specialist tacit knowledge is needed because such media are designed with the layperson in mind and cannot depend on specialist knowledge. But popular books and the like begin to give the reader the impression that they understand something and they, in fact, do allow the reader to understand more than they can learn from a beer mat. One isn't, however, learning as much as one thinks from these popular treatments.

Primary source knowledge is even more seductive. It is knowledge that is gained by going to professional journals and struggling through them, or reading highly technical material on the Internet. It is hard for the layperson to read this technical literature but not impossible. For example, it is possible to 'get the drift' of a physics paper in a technical journal without understanding the equations. Perhaps because of the 'degree of technical difficulty', as Olympic high-board diving judges put it, if one achieves some grasp of such a technical paper one tends to feel that one has reached into the heartland of the science. But it is not so. There is a huge gulf between this level of understanding, which rests solely on ubiquitous expertise, and the next one up, 'interactional expertise', that depends on acquiring specialist tacit knowledge. That gulf is indicated by the heavily marked divider in table 2.1. We will return to it in chapter 3.

Interactional expertise, the remaining category in line 2, is a relatively new notion – it seems not to have come to light until recently.[27] Interactional expertise is acquired by engaging in the spoken discourse of an expert community to the point of fluency but without participating in the practical activities or deliberately contributing to those activities. Thus, one can be an interactional expert in a specialist domain without being a 'contributory expert' in that domain (though all contributory experts are also interactional experts since they have to know the specialist language associated with the domain to which they contribute).

At first sight, interactional expertise seems a very 'thin' kind of expertise – 'talking the talk' but not 'walking the walk'. But interactional expertise is very important and very 'rich'; it is the key to much of what we do. First, it takes a very long time to acquire interactional expertise and when one has properly acquired it one should be able to pass as an expert in a test involving spoken fluency. For example, I have spent decades studying the sociology of the detection of gravitational waves and hanging around with the gravitational-wave physics community. I certainly cannot 'do' gravitational-wave physics but I decided to expose myself to a test that would indicate if I had a grasp of the tacit knowledge belonging to the specialists. In the test a gravitational-wave physicist asked technical questions of another gravitational-wave physicist and of me: we were told to answer quickly without looking anything up. The anonymized dialogue was then sent to nine other gravitational-wave physicists and they were asked to identify the real gravitational-wave physicist and Collins the pretender (they all knew me and that I was one of the contenders). Seven said they could not tell the difference and two identified me as the real physicist. The result was written up in the scientific journal *Nature* under the heading 'Sociologist Fools Physics Judges'.[28] Here is an example of a question and the two answers, one from me and one from the real physicist. It can be seen in table 2.2 that the acquisition of interactional expertise is not a trivial thing.

But interactional expertise is far more important than this experiment alone indicates. There are around

Table 2.2 Question and answers concerning GW physics[29]

A theorist tells you that she has come up with a theory in which a circular ring of particles are displaced by GW so that the circular shape remains the same but the size oscillates about a mean size. Would it be possible to measure this effect using a laser interferometer?

Yes, but you should analyse the sum of the strains in the two arms, rather than the difference (which is what's normally done). In fact, you don't even need two arms of an interferometer to detect GWs, provided you can measure the round-trip light travel time along a single arm accurately enough to detect small changes in its length.	It depends on the direction of the source. There will be no detectable signal if the source lies in any direction on the plane which passes through the centre station and bisects the angle of the two arms. Otherwise there will be a signal, maximized when the source lies on a line obtained by producing one or other of the two arms.

---

a thousand physicists working in the international, billion-dollar field of gravitational-wave detection. Each of them belongs to a sub-specialism within the area, such as working out the wave forms that will be emitted by various cosmic catastrophes, designing the lasers that are used in the laser-interferometer detectors, working on the principles of the detection apparatus, designing and building the exquisite vacuum system, mirror suspensions, and seismic isolation system, and so on and so on. In the main, no person from one subgroup could step in and do the work of a person from another subgroup – at least not without a long apprenticeship. If that were not so, they would not be specialists. And yet all these people have to coordinate their work. The way they coordinate their work is by sharing a common language which they learn when they attend the many international conferences that are part of their job, and by visiting and spending time at each other's laboratories. What they are doing is acquiring interactional expertise in each other's specialties. So, to make the point in a stark way, in principle, the only gravitational-wave-related expertise that someone who works in the specialism has, over and beyond an interactional expert like me, is contributory expertise in one narrow specialty. Otherwise it is interactional expertise all the way down, whether one is looking at one of the contributory experts or looking at Collins (or looking at the top managers who make the major technical decisions while not working in any one of the practical specialties). Thought about this way, it becomes a little less surprising that after decades hanging around with these scientists I could answer their technical questions.

Think about it some more and we realize that inter-actional expertise is the key to most of what happens in science. As mentioned, the managers of large scientific projects such as this one mostly work with inter-actional expertise.[30] The much-hyped 'peer-review' process for the vetting of journal articles and the award of grants works via interactional expertise because it will almost never be the case that the reviewers have actually done work that is identical to that which they are reviewing. And I can vouch for the fact that when physicists talk to each other, or when they sit on decision-making committees, they are not doing calculations and they are not doing experiments. They are talking the kind of talk that I learned to talk from hanging around: that is what feeds into their decision-making. Even learning to become a contributory expert in a narrow technical domain is mostly a matter of acquiring interactional expertise because it is through talk that one learns how to act in practical matters – one learns what one is supposed to take notice of and what one is supposed to ignore, what one should 'see' and what one should not see. When one engages in a practical activity which has, when thought about carefully, an indefinite number of variables and can be assembled into an indefinite number of patterns, it is mostly words that teach what is to be seen as the salient pattern. The world as it first impacts on the outermost layers of our senses is confused and featureless, but we use talk with others to learn how to break it up into the discrete objects that become the currency of the community of experts.

Interactional expertise provides both a hope and a warning to those who aspire to gain a substantive expertise in a specialist domain without going through the full process of apprenticeship. The hope is that not everyone who possesses some specialist expertise has to be a contributory expert; in fact most specialists are not acting as contributory experts most of the time; they are acting as interactional experts, using expertise that comes from what they have learned to say, not what they have learned to do. Bear in mind, however, that acquiring interactional expertise is not a trivial accomplishment – it is a substantive expertise. Interactional expertise is, then, something quite different from what is disparagingly referred to as 'talking the talk'. To acquire interactional expertise is to learn to *walk the talk*, as one might say: to acquire interactional expertise is to learn sufficient fluency in the talk as to be able to make the same judgements as any other expert, and that takes long immersion in the community.

What the idea of interactional expertise means is that there are a few more people who can legitimately be drawn into the fold of expert decision-making, *as technical experts*, than might once have been believed. In the past, a specialist expert was thought of as a contributory expert or nothing; now a specialist expert can be something just a little less – an interactional expert. But this does not widen the number of experts very much because there are very few interactional experts who are not also contributory experts. The idea of interactional expertise certainly does not justify the claim that we are all scientific experts now.

Meta-expertises

Row 3 of table 2.1 is a little more complex; this row, of course, is all about choosing between experts and their expertises, not executing expertises – though choosing between experts is itself an expertise and can be done well or badly.

One can try to choose between experts by checking their qualifications, or their experience, or their track records.[31] This is not easy: one has to know what a qualification or title really means, and to learn about a professional's experience or track record of success seems like a job for another kind of specialist. In any case, these criteria are unreliable. One thing that happens to scientists and other technical people is that they slip in and out of favour; their knowledge becomes out of date and, even if they have a good track record, each new problem presents new challenges which can be so unlike those that have been faced before that track record means nothing. Experts, as one might say, have a tendency to be 'fighting the last war' – the one they understood. Later in the book the controversy over the distribution of anti-retroviral drugs in South Africa will be discussed and we will see that one of the principal scientists involved was wonderfully qualified, had great experience and a good track record – the most prestigious scientist of them all – but when it came to judging his contribution for the purposes of policy, his qualifications and experience were misleading.

Setting qualification aside, and starting on the left of row 3 of table 2.1, is the expertise we use from day to day when we buy a used car or decide whether a politician seems honest or shifty. Everyone has to make such decisions so this is called 'ubiquitous discrimination': all of us discriminate between salespersons and so forth. Of course, it is a real skill as is shown by the regular stories of country bumpkins, or the very young, coming to the city to be fleeced. John Voight's naive character in the film *Midnight Cowboy*, visiting New York for the first time, could not discriminate between a prostitute and a client for his sexual favours though it was obvious to the sniggering cinema audience. Country bumpkins are a source of amusement because they have not learned the ubiquitous discrimination you need to survive in the city. But ubiquitous discrimination, even though it is hard to acquire, is not very reliable. If it was reliable there would have been no Watergate because everyone would have been able to see through Richard Nixon from the outset.

Moving one cell to the right in table 2.1, 'local discrimination' is what we can use if we know something particular about the persons we are choosing between in virtue of 'inside information' – perhaps we know others who have worked with them, or we live in their locality and have seen how they have acted at first hand. This is a strange kind of 'expertise' because it is based on local and specific information, but it is very important and can form the basis for justified 'whistle-blowing'. Whistle-blowing is a topic covered separately in chapter 4. Local discrimination is much more reliable than ubiquitous discrimination. The importance of local discrimination is indicated by the use of bold font in table 2.1.

Where experts disagree, for example, climate change experts or medical experts, one can see that citizens are forced to choose between them, usually on the basis of nothing but ubiquitous or local discrimination.

Climategate caused a fuss because suddenly we all thought that our discriminatory abilities had been hugely enhanced. We had been given local information about the Climatic Research Unit; whereas before we were struggling to think about who was right using only our ubiquitous discrimination, we now had the ability to exercise the much more reliable local discrimination. This seemed like a free gift and some immediately concluded that the global-warming argument and its proponents should not be trusted.

Both ubiquitous and local discrimination are referred to (in table 2.1) as 'transmuted expertises'. The alchemists' dream was to transmute the elements – to turn lead into gold – and both types of discrimination take the lead of judgement about persons and transform it into the gold of the kind of technical choice that is normally the preserve of specialists.

Moving to the right in row 3 of table 2.1, there are three non-transmuted expertises. They are called non-transmuted because they involve using expertise that is not primarily about other people but is a substantive technical expertise. Starting on the right is 'referred expertise', which is expertise from one technical domain being used in another technical domain. The idea of referred expertise is hard to apply and there is no need to worry about it for the purpose of this argument.

The next rightmost meta-expertise is 'downward discrimination'. Downward discrimination happens when a specialist makes a judgement about the quality of the claims of someone who is less expert in the same domain. The trouble with this expertise is that we all think we know better than the other person even when we do not, so who is to say that the expertise is being properly exercised in the downward direction? Who is to say who is right? It is very hard to answer this, though even a minor expert can recognize unrecognized downward discrimination. For example, in the UK in 2003 there was a public inquiry over genetically modified crops. Almost anyone with a smattering of technical knowledge could tell that one expressed anxiety, that the crops would be radioactive because radioactive markers were used in the manipulation of the genes, was without foundation. And yet the person who made the claim was probably unimpressed by the rebuttal – they thought they knew better and that the experts 'would say that, wouldn't they'. So downward discrimination plays a part in the relationship between scientists, technologists and the public, and it will show up again in chapter 3.

Technical connoisseurship, on the other hand, is regularly exercised by all those who employ craftspersons to do work on their houses. Though we may not be plumbers or tilers ourselves, we need to decide whether the plumbing work has been done properly and whether the tiles have been laid to an adequate standard. This is a skill that we pick up as we move around among tiles and pipes in our everyday world and is a ubiquitous expertise. Of course, when we want to be assured that a really high standard has been reached we may choose to employ a professional, such as an architect, to make the judgement.

We are now much better placed to talk about experts because we have a far richer notion of expertise and its types. It is no longer a matter of 'us and them' – public versus professionals. We can now ask what kind of expertise this group or that group claims to possess and what it actually does possess. Is it specialist expertise? If it is, how did members of this group acquire it and to what extent are they in touch with the culture of the professionals? Is it meta-expertise and, if 'yes', is it ubiquitous or local? If local, what special sources of knowledge have been tapped?

THREE

Citizen Sceptics

With the deeper and richer understanding of the nature of expertise discussed in chapter 2, we can take another look at some of the unfortunate events that have confronted the citizen. Let us start with Climategate.

The fascinating thing about Climategate is that it revealed to the general public a secret known only to a few – how science works from day to day. For many decades, the only people who knew this were experienced scientists who were capable of reflecting on their practices. In the 1970s, Wave 2 of science studies began to reveal the secret to social scientists. I remember from the early 1970s, the lascivious delight I took in simply quoting scientists' words when they were criticizing others. Just quoting these words was opening up a new world because no one outside science knew that scientists talked or thought that way. Wave 1 of science studies was all reverence for the most perfect form of knowledge there could be and the priest-like abilities of those who knew how to commune with nature; it was all white coats and 'aha's! Wave 2 found the priests profaning and blaspheming and not so holy after all. That this was so was a consequence of such things as the experimenter's regress: if assessing the value of the scientist was part and parcel of assessing the value of his or her findings, then inevitably a language appropriate to the assessment of people as well as nature would be spoken, and it wouldn't always be polite.

It is not easy to take account of this aspect of the scientists' world while still holding on to the idea that there is a discrete and special set of practices and values identifiable as 'science'. It took a long time for the apparent clash between science as it used to be described and the science that Wave 2 discovered to 'shake out'. What the Wave 2 supporters were describing was often interpreted as an attack on science, not least by scientists, and it is no surprise that it was. Only decades later are the researchers and their targets working out how to describe science accurately while still admiring it and what it stands for. The trick that has to be learned is to treat science as special without telling fairy stories about it. This is the project called 'Wave 3' of science studies.

One way to reconcile the special nature of science with an accurate portrayal of the scientists' world is to describe and analyse the skills, experience and expertises of its practitioners. That is what we have been leading up to. Another and perhaps ultimately more important way to minimize the impact of the profanity of the priests is to put the evidence together differently. In their moment-to-moment actions scientists may act in a mundane way but the *spirit of science* is still 'divine'. It is what science is aiming for that matters and so long as this ideal drives the actions of most scientists most of the time, then Wave 1 of science studies was not so far out in its respect for the norms and values of science. That, indeed, is why we are so offended when we find that scientists have been bribed by the tobacco and oil companies to produce bespoke findings tailored to fit the company's financial interests. We know, immediately, that something wrong is taking place and this shows that we know how science ought to be and mostly is. Contrast this with the arena of politics where there is nothing unusual at all about lobby groups being funded by commercial companies – that is simply how the political world works. If it ever comes to be the case that it is accepted as a matter of course that it is how the scientific world works, then science will cease to exist. We will return to these points in chapter 4.

The main impact of Climategate came from the fact that it was the public's first glimpse of the profane world of science and the public took it that the scientists under the spotlight were acting untypically – as discussed in the last chapter, the public thought it had insight into something *local* to do with the Climatic Research Institute. In one sense it was gratifying that the public were offended: it showed, once more, that the public understands what science is meant to be. But those who understood Wave 2 knew that the climate scientists were not acting untypically – what was being exposed was not something special and local but 'business as usual' across the whole scientific world. It is just that the Wave-2ers, or at least some of them, had learned, after a struggle, to reconcile this business as usual with the traditional idea of science.

It has to be said that not every reading of Climategate in the mistaken, local, way is an innocent one. It was strongly in the climate change sceptics' interest to have the Climategate emails interpreted as a breach of scientific objectivity – as essentially a set of political acts. The problem – distance lends enchantment – is represented in 'the target diagram' (figure 3.1). The target diagram shows the scientists doing the work – the 'bullseye' – and the rings of people who discuss it, evaluate it and report it from ever increasing distances.

In the middle of any scientific dispute is a 'core-set' of specialists – these are the people who actually do the experiments, build the theories, and meet together to argue at conferences. In the early days of a debate over something like gravitational-wave detection, the

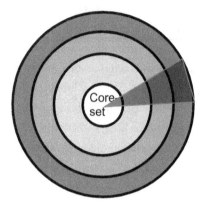

Figure 3.1 The target diagram

number of scientists in a core-set can be no more than half a dozen or so. In the case of gravitational-wave physics, however, and this is what happens where the science is important, what the core-set did was being reported and discussed in the outer rings by hundreds of their fellow scientists, by funders and policymakers, by journalists and, to some extent, by the public at large.

The key insight is that what happens inside the core-set is hugely complicated. In the early 1970s, every waking moment of the scientists locked in dispute about whether gravitational waves had actually been detected was filled with calculations, arguments, measurements, judgements of others' capabilities and so on. How could it be otherwise? That is what a committed professional's life is like. Climategate is just a glimpse behind the scenes of a small part of the blooming, buzzing hive of activity – after all, there were only a few emails.[32]

What it means to be a 'specialist' is to be in there with all these goings on, twenty-four hours a day. To be a non-specialist is not to be in there. If you are outside, things inevitably become simplified. The 'bandwidth' of the channel to the outside is too narrow to carry all the nuanced information about what is happening inside the core-set and it would be a full-time occupation to absorb it: to swim in its water you would have to become a professional or semi-professional yourself. The outcome in respect of those outer rings is that *distance lends enchantment*. What is nuanced and unclear to those inside the core-set becomes, paradoxically, sharp and clear to those outside it. Knowledge roughly follows a 'direct square law' – bizarrely, as it travels further, it becomes rapidly stronger because all the uncertainties get lost. So people outside the core are much more certain of what is going on than the people inside who are making it happen![33] Sometimes the outside rings are more certain in a positive way but, where there is disagreement (the shaded segment), those who chose the minority view will be more certain about that negative position. An untidy set of doubts in the centre becomes a competing and polarized set of certainties as the distance increases. For example, see how certain both the climate-warming sceptics and the climate-warming enthusiasts are of their respective positions; they are more certain of what has been found out than the climate scientists themselves! The scientists may be pretty sure they are right but they do not possess the almost religious certainty of either the sceptics or the true believers. A messy science like climate change cannot deliver these quasi-religious convictions.

To repeat, we know from Climategate, and from decades of careful examination of scientific practice by social scientists, that the difference between the inside and the outside is not what we once thought it was. We know that scientists' activities inside the core-set look pretty ordinary in many respects once you get close enough. For these reasons, we know that we can never go back to the 1950s where the pronouncement of any scientist in a white coat was taken to be authoritative, not only on the science, but also on any policy-related issue. We know that pronouncements on even the science are no longer authoritative and that democratic politics always trumps scientific conclusions. We know that it is better that people know more about the processes of science and understand how this conclusion or that will affect their lives. Nevertheless, it must not be forgotten that the inside and outside of science are different in two important ways. First, as explained, there is all that detail and nuance trapped in the middle and invisible from the outside, and, second, there is a different style of argument within the core because of that detail and nuance and because of the value system of science.

We know the value system of science is often honoured in the breach but, as argued, and as will be reiterated nearer the conclusion, it still underpins its distinctiveness. Very roughly, it means that, mostly, inside the core, you are trying to get to the collective truth of the matter and this means that if you want to do a good job you must start by trying to understand and fairly represent your opponent's position. You have to do this if you want to convince your opponent as well as yourself that you are right. And you have some grounds, albeit vestigial and often forlorn grounds, to hope that you can convince your opponent with argument so long as you start from his or her position. This may just happen, and does happen on extremely rare occasions – but that is better than never and is enough to make it worth the effort. Convincing others in the core is just possible because everyone inside knows about the nuances and the doubts. Outside, there is no such hope because no one knows enough of the detail and the doubts to get themselves out of their polarized positions, and that is why disagreements turn into 'campaigns' rather than debates.

The distinction between a campaign and a debate may seem subtle but it is not; scientists immediately know when their opponents have ceased to play by the rules and instead of taking their opponents' arguments seriously they are ignoring them or caricaturing them and 'playing to the audience'. At that point the scientist is directing argument, not at the core, but outside towards the public. That is 'science war' not science debate. Therefore, if one wants to preserve the thing called 'science', as a distinctive way of making knowledge, one cannot mix the outer rings of the target diagram too thoroughly with the core – the appropriate ways of going on are too different.

As explained, those who innocently misinterpreted Climategate felt that they had gained some special knowledge about the scientists who put forward the anthropogenic climate change theory that showed them to be less trustworthy than other scientists. They 'transmuted' this political insight into a technical conclusion: that the Earth was not being warmed by human activity. But the specialness of the profanity of the Climategate scientists was illusory. This is not to say that the anthropogenic warming thesis is right – I am not doing science in this book and I know no more about climatology than any other citizen – it is only to say that what happened at Climategate does not show the anthropogenic warming thesis is wrong or even suspect; in fact, Climategate does not bear on it in any way.

There is a second lesson to be learned from Climategate that also applies more widely. Consider the first two of Jim Sensenbrenner's list of suspect emails (Introduction, p. 12):

From Kevin Trenberth:

The fact is that we can't account for the lack of warming at the moment and it is a travesty that we can't. The CERES data . . . shows there should be even more warming: but the data are surely wrong. Our observing system is inadequate.

From Phil Jones:

I've just completed Mike's Nature trick of adding in the real temps to each series for the last 20 years (i.e., from 1981 onwards) and from 1961 for Keith's to hide the decline.

Subsequent analysis has shown that neither of these comments is anything like as revealing as it appears. Taken out of context, the first seems to suggest that the climate change scientists are peddling inadequate or incorrect data; the second suggests that trickery was being employed to hide a decline in temperatures.

Here is Trenberth's response to the first:

> In my case, one cherry-picked email quote has gone viral and at last check it was featured in over 107,000 items (in Google). Here is the quote: *'The fact is that we can't account for the lack of warming at the moment and it is a travesty that we can't.'* It is amazing to see this particular quote lambasted so often. It stems from a paper I published this year bemoaning our inability to effectively monitor the energy flows associated with short-term climate variability. It is quite clear from the paper that I was not questioning the link between anthropogenic greenhouse gas emissions and warming, or even suggesting that recent temperatures are unusual in the context of short-term natural variability.[34]

As to the second quotation, taken in context it turns out that the 'trick' that was mentioned was scientists' normal way of talking about a 'neat trick' for accomplishing some technical transformation, while the 'decline' was in tree rings and not temperature.[35]

Here one can begin to see how difficult it is to understand meaning from text alone. Just as in any use of language, to know the meaning of words one has to know how they are normally used in the kinds of contexts in which they appear; words are not numbers, while even numbers are not unambiguous out of context. In sum, these first couple of Climategate quotations indicate why it is hard to make useful specialist judgements without possessing the tacit knowledge that comes with being a member of the expert community.

Science as a collective activity and tacit knowledge

There are many different elements of the tacit knowledge that allows a specialist community to make sense of the literature, published or informal. One can see why by starting with a deep point: *science is a collective activity*. It is hard to see this because science is also an individualistic activity: our favourite myths concern individual scientists who have stood up to the community and turned out to be right in the long term. In the 1600s, Galileo had to take on the Catholic Church to promote the heliocentric theory; he remained under house arrest for the rest of his life. Things move faster nowadays and the penalties for dissent are not so drastic (burning at the stake is no longer on the menu), but the more things change the more, in essence, they remain the same. Thus, in the 1980s, Barry Marshall and Robin Warren had to take on the medical profession when they claimed that a virus caused stomach ulcers; they had their papers rejected and it took them years to convince a sceptical medical community (they were awarded the Nobel Prize in 2005). And yet we consider that these individuals were vindicated because what they believed they had shown is now the collective view of scientists. It is the community of scientists who decide on what we are all going to accept as the truth of the matter; we, as lay-individuals, have no technical grounds on which to decide that Marshall and Warren are right before the scientific community has done its work, however slowly they go about it.[36]

In any case, scientific heroes – and if science is to remain science we need individual heroes – are great discoverers, not ordinary people trying to find their way through a disputed territory to make an everyday decision. No non-specialist should be able to get credit for aligning themselves, on technical grounds, with a maverick scientist who may, one day, turn out to be right. There are huge numbers of mavericks and to align oneself with any one of them is nearly always simply a matter of taking a gamble with very, very, long odds. It makes no sense for ordinary people to take such a gamble. It makes very little sense for a specialist to take such a gamble unless they have some privileged source of information – either something to do with the science itself or the fact that the maverick scientists' work is being treated unfairly. But what specialists tend to know about mavericks, that it is hard for ordinary people to know, is the history of how their work has been treated in the scientific community.

I can give a graphic example of this from my own experience in the gravitational-wave community. The pioneer of terrestrial gravitational-wave detection was the late Joseph Weber, universally known as 'Joe Web[b] er'. Weber built the first serious detector and without him we almost certainly would not have the billion-dollar international effort we see today. In the late 1960s and early 1970s, Weber said he was detecting the waves with his relatively cheap and insensitive detector. The results were disputed and, by around 1975, Weber was no longer believed by the community of gravitational-wave scientists.[37] After 1975, however, Weber continued to press his claims.

In 1996, Weber published a paper in the physics journal *Il Nuovo Cimento*, in which he claimed that the waves he believed he had seen two decades earlier had been correlated with another strange cosmic phenomenon – gamma-ray bursts. He said the correlation was supported by a level of statistical significance of 3.6 standard deviations – which is not quite the 5-standard deviations that physicists look for in the case of new discovery but high enough to have given rise to a flurry of excitement, new directions of research, and, if confirmed, a Nobel Prize.

In 1996, being well embedded in the gravitational-wave community I was able to ask many of the physicists what they made of this latest claim by Weber. I discovered to my astonishment that the only person who had even read the paper, apart from Weber and the editorial team, was myself. None of the gravitational-wave physicists had even looked at it. And this was because Joe Weber's credibility had long gone.

It is not impossible that Weber was right and his old waves were really there. They could have been generated by some strong local source and are no longer visible to the much more sensitive detectors because their source has gone. It is possible that they were correlated with gamma-ray bursts because gamma-ray bursts are not what astronomers and astrophysicists think they are, which is the symptom of huge explosions in vastly distant galaxies, but relatively local generators of gravitational waves. All this is possible in logic. But how can any non-specialist have a chance of supporting such a claim against the graphically expressed opinion of several specialist scientific groups? The specialists say all those possibilities are crazy and my interactional expertise tells me why. To support Weber's claim against the mainstream would be impossible to justify unless one had some very esoteric knowledge indeed or could demonstrate the existence of a huge conspiracy.

But notice that it is possible for an outsider to read Joe Weber's 1996 paper and from it gain 'primary source knowledge' – the third category from the left of the specialist expertise row of table 2.1. I can promise such lay readers that if they teach themselves a bit of elementary statistics and persevere with reading the paper, they will find it utterly convincing. Scientific papers are written to be utterly convincing; over the centuries their special language and style has been developed to make them read convincingly. Try reading it. The reference is: Weber, Joseph, and B. Radak. 1996. 'Search for Correlations of Gamma-Ray Bursts with Gravitational-Radiation Antenna Pulses'. *Il Nuovo Cimento* B9 1119 (6): 687–92!

The only way to know that Weber's paper is not to be read in the way it is written is to be a member of the 'oral culture' of the relevant specialist community. This is the only way the reader can know that the consensus of the scientific community is that the paper is to be ignored. It is the only way for you to acquire that sure sense that, by 1996, the pioneering scientist who wrote it had published so many results that were rejected as to reduce his credibility as a novelty-maker in respect of gravitational waves to zero. There is nothing in the paper that will demonstrate this. Nor, to repeat, could there be, because Weber might – just might – have been right. But for an outsider to bet that he was right would be crazy. This case is very important because, as we shall see, similar cases have had grave consequences.

What this case illustrates is that in the specialist expertise line of table 2.1 there is a huge gulf between primary source knowledge and interactional expertise. A person with interactional expertise in gravitational-wave physics will understand how to assess Weber's 1996 paper; a person without it will see no distinction between that paper and any other gravitational-wave physics paper. So the vertical line that separates primary source knowledge from interactional expertise is the most important division in the whole book, and that is why it is heavily emphasized in table 2.1.

Going back to the Climategate emails, to know what they mean one would need, at least, interactional expertise. One cannot, then, know what they mean without being an insider. Strangely, this point is missed by at least some of the insiders. Consider the view of Mike

Hulme and Jerry Ravetz. Mike Hulme is professor of climate change in the School of Environmental Sciences at the University of East Anglia – the origin of Climategate – while Jerry Ravetz is a scientist turned philosopher and social analyst of science. They insist, in response to Climategate, that scientists must 'show their working'. They write: 'To be validated, knowledge must also be subject to the scrutiny of an extended community of citizens who have legitimate stakes in the significance of what is being claimed . . . in the new century of digital communication and an active citizenry, the very practices of scientific enquiry must also be publicly owned.'[38]

But it is impossible for scientists to 'show their working' and impossible for science to be publicly owned. For science to be publicly owned, everyone would have to become at least an interactional expert in all the sciences in which they have an interest. To become an interactional expert requires an apprenticeship. And, as explained, the huge amount of activity that takes place in the core-set cannot be 'shown' without actually living it. Without this, how is it going to be understood that scientific judgements, and this includes the much vaunted peer-review process for the publication of journal articles and the distribution of research resources, are as much a matter of judging scientists as judging their work? The 'working' that would have to be shown would, for example, include the equivalent of the fact that the 1996 *Nuovo Cimento* paper of Weber was not read because by that time scientists had decided to ignore Weber. Because the logic of science is never closed – anything is possible – if science is ever to move on it has to be driven by what is appropriate 'for all practical purposes'. Weber's 1996 paper might have been correct but, for all practical purposes, it was dead from the moment the print dried on the paper. It is difficult to 'show' this in a convincing way that does not lead to immediate dispute, but it can be understood by experts – by both scientists in the field and by social scientists who understand the processes of science. This kind of knowledge has enormous practical importance.

Unhappily, the practical importance can be illustrated by the tragedy of anti-retroviral drugs in South Africa. The South African government decided not to distribute anti-retroviral drugs to, among other groups, pregnant women, even though it was said it would prevent transmission of HIV to their unborn babies. President Thabo Mbeki justified this decision with a speech before parliament. In 1999 he spoke as follows to the second chamber:

> There . . . exists a large volume of scientific literature alleging that, among other things, the toxicity of this drug [the anti-retroviral AZT] is such that it is in fact a danger to health. . . . To understand this matter better, I would urge the Honourable Members of the National Council to access the huge volume of literature on this matter available on the Internet, so that all of us can approach this issue from the same base of information.

Mbeki was right, and he was 'showing his working'. But what he did not understand, or decided not to make clear, was that this material was being promulgated by scientists in a Weber-like position: scientists who could no longer publish their views in the mainstream journals. There had, at one time, been a serious scientific debate about the value of AZT but it was dead 'for all practical purposes'. To know this, however, requires more than reading the journals – what has to be understood is the process by which the scientists pressing the idea of the dangers and inefficacy of AZT had come to be marginalized. These marginalized scientists included, perhaps, the most famous and experienced scientists in the whole debate, Peter Duesberg and Nobel Laureate Kary Mullis. In spite of these scientists' fame and achievements, by the time of Mbeki's speech they were ignored within the mainstream scientific community.

What was going on was not a real scientific controversy but what we can call a *counterfeit scientific controversy*. In a case like this, only time, and it has to be a very long time, can tell who was truly right, but to make the *appropriate* judgement, or the *best* judgement, one needs to know the social process of science and know where that process stands in the particular case; there is simply nothing better to go on.

Mbeki, by presenting this counterfeit controversy as a real scientific controversy, was actually disempowering the South African political process. It is not possible for ordinary people to establish, in the face of their President, that there was no real scientific controversy, so there were no politics in which they could engage – they had to accept the verdict of the science as presented by Mbeki. If Mbeki had said, as he was perfectly entitled to, that the mainstream scientific community believes that AZT will save tens of thousands of unborn children from acquiring HIV from their mothers, but that South Africa is a poor country and cannot afford the drugs, nor does it wish to fall under the thrall of Western pharmaceutical companies, and does not wish to project the image of a promiscuous, disease-ridden land, and therefore the government has chosen not to distribute the drugs, he would have left his people the chance of expressing their opinion of that political choice via the ballot box; as it was, the option was taken from them.

It is said that the non-distribution of AZT to pregnant women resulted in tens of thousands of unnecessary cases of HIV. *But if it had not led to even a single extra case* the argument would still hold. There was no real scientific controversy in 1999 and the best decision had to take account of the current scientific view: AZT was an efficacious and relatively non-toxic drug. That scientific view should not bind the policymaker but it should have been taken into account and fairly presented.[39]

The conclusion of this chapter is that citizens must be careful of being too sceptical when they discover that much of the day-to-day activity of science has a mundane look to it and that scientists when they argue with each other use a vocabulary that is much like that of ordinary people. This does not mean that their aspirations are no different from those of ordinary people and that their judgements are not directed towards finding the truth of the matter rather than, as with politics, driven by self-interest. It is just that finding the truth of the matter requires frequent excursions into ordinary-looking debate. On the other hand, when the debate is not so ordinary-looking – when there are maverick claims in the specialist journals which are, from the outside, indistinguishable from any other scientific claim – it is the specialist ability of the scientific community to know how they stand in the world of credibility that is important. Ironically, scientists at these points have to rely on ordinary-looking debate, but it is 'specialist ordinary-looking debate', to which the ordinary citizen has no access. There are, then, huge dangers in ordinary citizens trying to make judgements about what is happening inside the scientific community from their inevitably distanced and unavoidably more polarized position outside it.

FOUR

Citizen Whistle-blowers

I have argued that ordinary people are not in a position to make technical decisions in the kinds of cases described above; nor are they in a position to make sound judgements in respect of the social process of science. In both cases, at least interactional expertise is required and ordinary people do not have it. But there is a kind of substantive judgement that ordinary people can make and it lies within the left-hand section of the 'meta-expertise' line of table 2.1; it emerges from properly applied local discrimination.

What is known as 'whistle-blowing' is an example of the use of this kind of meta-expertise. Non-specialists may have enough knowledge of a local situation to understand that the normal scientific process is being distorted. To know this does not require specialist knowledge of the science, only the most general knowledge of what kind of thing science is. That is enough when such things are discovered as that the tobacco companies are generously funding scientists who then go to conferences and cast doubt on the link between tobacco and cancer: it is enough to understand that the tobacco companies are trying to create a *counterfeit scientific controversy* of the same type that Mbeki was knowingly, or unknowingly, invoking.[40] This, then, is the kind of occasion when local discrimination is legitimately and usefully deployed and a non-scientific expert can generate transmuted expertise – expertise about science generated by non-scientific means. One has to be careful, of course, that one does not make the kind of mistake that can be made over Climategate. One has to be careful of mistaking the ordinary processes of science for distortions of the scientific process; these are distortions only when compared to the idealized Wave 1 model. But taking that care does not mean it is impossible for the citizen to spot when scientists are being paid to deliver specific conclusions or the like.

Vaccine protestors

Consider, in the light of the richer analysis of expertise, anti-vaccination campaigners – those who campaign against the vaccination policies of nations. Let us agree that they are not 'default experts' – that is to say, it is not that there is no substantive expertise to vaccination. Let us agree that, for example, vaccination has eliminated smallpox and polio from the Western world, though polio is still prevalent in, for example, parts of Nigeria where they refuse vaccination with oral polio vaccine. We know that at least some of the vaccine protestors think that these diseases have been eliminated not by vaccination but increased standards of hygiene, and we know that some think the natural process of catching a disease and fighting it off is better than trying to prevent it, but we have to start somewhere. That 'somewhere' is that some vaccines have worked so there is some substance to vaccine science and it is not, in itself, a bad thing to use vaccines to combat disease, even though it may be a bad thing to enforce the use of some particular vaccines. With that starting point, we can look at contemporary debates.

In the late 1990s and early 2000s, there was a revolt against the use of mumps, measles and rubella (MMR) vaccine in the UK. It was not a revolt against vaccine in general since the person who started the trouble recommended continuing treatment but using single rather than combined vaccines. This person was a medical doctor and researcher, Andrew Wakefield, who said that he had examined half a dozen autistic children and found measles virus in their gut. At a 1998 press conference, Wakefield made the claim that the autism could have been caused by the MMR vaccine and recommended three separate vaccinations instead of the combined MMR. The press picked up on this and gave huge publicity to parents who said that their children had shown the first symptoms of autism shortly after an MMR jab. In the newspaper stories, these parents' accounts were 'balanced' with the reports of epidemiologists that showed there was no increase in autism rates in countries that had newly introduced the MMR vaccine, so there was no evidence for any correlation between the two. Nevertheless, the resistance to MMR grew and parents who could afford it started to purchase single vaccine shots even though there was no evidence that the measles virus in a single shot would work any differently from the measles virus in the MMR shot. Certain members of the social science community demanded more research on the matter, while members of the opposition Conservative Party championed 'parents' right to choose' and demanded that the Labour Government abandon MMR and offer single shots, even though vaccination experts considered that the single shots were less effective. The government stood firm but the revolt was such that the UK now has a minor measles epidemic; herd immunity was sacrificed when the measles vaccination rate dropped below a certain level. A small number of children are now dying or being maimed by measles, while the incidence of autism has not changed.[41] It subsequently turned out that Wakefield was in the pay of companies that would benefit from a shift to single vaccines and that he had also violated ethical procedures in the collection of evidence.[42] Let us not worry about these subsequent revelations – we are no more interested in being wise after the event in this case than in the Thabo Mbeki case. Being wise after the event is too easy. The question is: upon what expertise was the revolt based at the time it was going on?

I now have to invoke my own expertise and, in particular, the fourth category from the left in the meta-expertise row of table 2.1: 'downward discrimination'. Downward discrimination is a dangerous category but, as intimated in the earlier discussion, 'even a minor expert can recognize certain kinds of mistakes even if they are not acknowledged by the party making them'. This, it will be recalled, was exemplified by the protestor at the GMO meeting who thought the use of a radioactive marker in the gene manipulation experiments would make the commercial crops radioactive. My qualifications are a training in science sufficient for me to carry out statistically analysed experiments in my own field (I hold a €2.26m European Research Council Advanced Grant to work on large-scale, statistically analysed, Imitation Game research), along with my interactional expertise in gravitational-wave physics – a statistically analysed science – and my sociologist's understanding of many debates and disputes in various other sciences. This is enough for me to say that the observation of half a dozen autistic children with measles virus in their gut could not establish a link between MMR and autism. In fact, it amounted to no evidence whatsoever on the relationship between MMR and autism. It was pretty well no evidence at all, but if it had borne upon anything in even the smallest way it would have been a relationship between measles virus and autism, not between MMR and autism. The strange thing, evident at the time, was that Wakefield, nevertheless, continued to recommend single-shot measles vaccination. The reports in the press, 'balancing' parents' views with those of epidemiologists, led to a *counterfeit scientific controversy* and seem to have arisen out of journalists' desire for a story in the absence of any attempt or ability to understand the science. There simply was no science on Wakefield's side and this was a shameful episode in newspaper history.

For the parents who reported that their children had become autistic as a result of the MMR jab, one has to feel a little more sympathy. There is no reason to expect ordinary people to understand even the elementary statistics involved. What they saw was a striking and horrible sequence of events with what seemed an irresistible logic: their child was given an MMR jab and shortly after their child showed symptoms of autism. In the face of tragedy we all search for a cause and will even settle for some wrong we committed in the past for which some god is punishing us. How could the parents see anything other than that the jab caused the autism? The psychological pressure to see it that way would be overwhelming. But a cooler analysis reveals that, typically, the symptoms of autism tend to reveal themselves around the recommended time for an MMR vaccination so it is a statistical inevitability that there would be a number of cases where the jab immediately preceded the onset of symptoms. There would be an equal number of cases where the onset of symptoms preceded the appointment for the jab but no one would think autism caused the appointment, whereas everyone would think the jab caused the autism. This explains the forcefulness of the supposed evidence as far as individuals are concerned, the only countervailing power being epidemiological analysis which showed in study after study that there was no connection between the jab and the autism. One can see how misplaced it was for the newspapers to present these two forms of evidence in a 'balanced' way.

But, of course, I am merely reporting what I read in the popular science journals. I am not an epidemiologist and can only report what others have told me. There is still room for me to be wrong given the meagre store of expertise on which I am basing my downward discrimination. I could be wrong if the reports of the safety of the MMR vaccine and the epidemiological evidence were fabricated as a result of conspiracy by the drug companies to maintain their profits (or a result of a conspiracy by the sinister agencies to deliberately damage the health of children, as is believed in respect of the Nigerian polio vaccine revolt). This, after all, would be equivalent to the kind of thing discovered by investigative journalism in the case of the tobacco companies.

The crucial point is that conspiracies are all that we are left with: however much parents feel they have the evidence of vaccination's dangers before their very eyes, they are wrong. The only kind of expertise left to us in such a case is *meta-expertise*. What kind of meta-expertise is it? Is it ubiquitous discrimination – the ability to see a shifty look in the eyes of doctors and epidemiologists or some such? Let us hope not because that is far too unreliable a basis for a decision of such magnitude.

Vaccination is a serious business because it affects not just the parents of the vaccinated child, and not just the children and grandchildren of the parents: it affects the lives of my child and your child. When celebrity Jenny McCarthy uses her personal experience of mothering an autistic child, and her media power, to convince other parents that vaccines are dangerous, she is affecting the lives of our children and grandchildren, not just her own. Vaccine is most effective when it results in the development of herd immunity and the eradication of a disease, in the way it has eradicated smallpox and polio in the West. But the development of herd immunity requires the maintenance of a certain high vaccination rate – around 90 per cent in the case of measles. Insofar as the anti-vaccine campaigners persuade parents not to vaccinate they are persuading them to expose your child to a future measles epidemic – or a pertussis epidemic or a polio epidemic. I remember polio from when I was a child: it was scary. Every now and again my class of eight-year-olds was warned to stay away from the public swimming pool. And who are the most likely to suffer? They are poor, undernourished children living in overcrowded conditions where disease spreads easily and bodies are less well equipped to fight when the disease strikes. They are children whose immune system has been damaged by illness or transplant surgery. And who are those most heavily involved in anti-vaccination media campaigns? They are the middle classes and the 'media savvy'. Anti-vaccination campaigns, *unless they are based on real danger*, are not an example of democracy at work except insofar as democracy is the rich and the strong undermining the poor and the weak.

How, then, can we judge whether these campaigns are based on real danger? We must not take the slightest notice of the technical claims made by the campaigners – they have no grounds on which to base technical claims – but must use our own meta-expertise to ask whether they have deployed their resources in such a way as to have uncovered a real conspiracy. Have they done the hard work necessary to uncover a real conspiracy in the way that was shown for the tobacco companies and can they demonstrate that they have done it? Consider this:

In April of 2008, CNN's Larry King hosted a show on the vaccine–autism 'debate' featuring Jenny McCarthy, a celebrity 'autism mom' promoting a book about her son Evan's 'recovery' from autism. McCarthy told King that she speaks to thousands of moms every weekend who relay the same experience: 'I came home, he had a fever, he stopped speaking, and then he became autistic.' 'It's time to start listening to parents who watched their children descend into autism after vaccination,' she urged, because 'parents' anecdotal information is science-based information'.[43]

But anecdotal evidence of this kind is not science-based information, it is at best isolated bits of data that are meaningless before they have been assembled and analysed.[44] If they are serious, Jenny McCarthy and her colleagues must do the hard investigative work of pinning down the conspiracy. This is where Hulme and Ravetz come into it: when the job is done the investigators should 'show their working', just as investigative journalists show their working. That is the kind of working we *can* understand because it is not the preserve of specialist communities but something close to the discrimination we use in our everyday lives when we recognize corruption. Nevertheless, this has to be done in the style of insiders' work, where the truth is being sought, not in the style of outsiders', polarized, campaigns. It is necessary for the investigators to show *how* the corruption of the doctors and epidemiologists has been uncovered, and show it in a way that will be (potentially) convincing to their opponents, not just those who are already determined to believe. It is a task that has to be done to the highest standard because otherwise we will be subject to ill-founded attacks on our children's and grandchildren's well-being, with a special danger to the poor and disadvantaged. We need, then, the equivalent of a Watergate investigation rather than the felt certainties of a parent, however emotionally convincing those feelings are.

Are we all experts now?

In this book the world of expertise as it applies to ordinary citizens has been divided into four categories: default expertise, ubiquitous expertise, specialist expertise and meta-expertise. We can now summarize what has been said in each case. Default expertise will be left to the end as it provides the most compelling case for us all to be treated as experts and the most difficult case to answer.

Ubiquitous expertise

We are all experts now because we are all immensely accomplished in our ubiquitous expertises – the expertises we use to live in our societies. Ubiquitous expertise is mostly overlooked in the analysis of expertise, though it has been rediscovered as we have tried and failed to make computers that act like humans. In the case of decision-making where there is a scientific dispute, however, this ubiquitous expertise cannot give rise to sound judgements. Using our ubiquitous expertise to read the literature may give us the feeling that we are acquiring deep knowledge, but it is an illusion because we have nothing to tell us what parts of the literature to take seriously and what to ignore; to know this, one needs interactional expertise.

Specialist expertise

One small and very unusual group of specialist experts are 'special interactional experts': people who, like the author of this book, acquire interactional expertise through occupying a strange role in which they immerse themselves in the discourse of a specialist community without fully participating in that community's expertise. According to the argument of this book, they are a newly discovered kind of contributor to scientific debates. In respect of science and technology, the largest numbers of these are probably found among science writers and high-level science journalists, though it is shocking to discover how few interactional experts there are among the journalists who write about science, at least for the British press. The science of economics is probably represented by the greatest number, while the MMR revolt shows just how badly health is served in spite of the fact that the largest amount of science journalism in our newspapers is devoted to health.

Almost everyone who works for a living has a specialist expertise: the expertise associated with the training or experience they gain in doing their specialist job. In that sense too, we are all experts now. But this provides no warrant for the claim 'we are all *scientific* experts now'. There are people, such as the sheep farmers on the post-Chernobyl Cumbrian fells or the farm workers spraying 245T herbicide, who are generally classed as laypersons but who have acquired some form of technically powerful substantive expertise through their work. Where that expertise is relevant, they can and should contribute to the specialist debate. It is disgraceful that the respect for qualifications and formal education has excluded them in the past. But by the time such people have acquired their specialist expertise they are no longer ordinary people and they are certainly not 'all of us'. These experts come in small numbers – they are just different kinds of elite. Those who live with chronic diseases also sometimes become experts on their own disease. The AIDS activists make up another interesting group: they reacted against the double-blind testing of anti-retroviral drugs. The longer they were engaged in the battle against the standard medical drug-testing regime, the more expertise they gained, in the end winning the respect of the medical researchers for their understanding. So we can say that another kind of group comprises dedicated and long-term activists who can acquire interactional expertise and, in rare cases, even a degree of contributory expertise.

Mostly our treatment of activists has leaned in the other direction. Activists who acquire their knowledge from the Internet or even from an independent reading of the primary sources are dangerous. Those laypersons who acquire 'primary source knowledge' gain the almost unavoidable impression that they have true expertise in the science but, impression or no, they do not have the interactional expertise, or 'specialist meta-expertise', that goes with being part of the oral community of the contributory specialists. The tragic consequences that can flow from mistaking one thing for the other are illustrated by Thabo Mbeki's argument for the non-distribution of AZT. Tragic on a small scale, but potentially just as disastrous as the South Africa case, are the anti-vaccination campaigns – *insofar as they are based on the campaigners' impression that they are in possession of technical knowledge.* Even while the number of deaths and disabilities that arise from these campaigns is currently low, numbering in the hundreds, for most of the readers of this book they are more immediate. A single child with polio or one who, unnecessarily, dies from or is maimed by measles, is as big a tragedy for the child and family as there can be.

Meta-expertise

But are the anti-vaccination campaigns baseless? Even though the campaigners do not have the technical knowledge to argue their case, they could have relevant meta-expertise. Let us start with the expertise we all possess – ubiquitous meta-expertise. That is the expertise required to choose between politicians and salespersons by reference to their demeanour and their readiness with answers. It is what Hulme and Ravetz are referring to when they say 'The public may not be able to follow radiation physics, but they can follow an argument; they may not be able to describe fluid dynamics using mathematics, but they can recognise evasiveness when they see it.'[45] Recognizing evasiveness is an element of ubiquitous meta-expertise.

Yet do we really believe ubiquitous meta-expertise can be a good enough foundation from which to mount a campaign that can damage generations to come by re-establishing eradicated diseases? Where did this idea come from? Look at the UK and the popularity of capital punishment in the face of endless expert testimony that it does not deter! Look at the USA and the gun lobby! Look at the admiring crowds as Adolph Hitler screamed his farragoes of utter nonsense – where was the ubiquitous meta-expertise? Ubiquitous meta-expertise is the only possibility when it comes to democratic elections; it is the very idea behind legal juries, who, nevertheless require special instructions from the judge and are sometimes not used when a case is highly technical, and it is all we have to go on when choosing between salespersons and products – though there it is manipulated by the advertising industry. But why rely on ubiquitous discrimination in serious matters when there are better things to go on?

A better resource, which still rests in the hands of ordinary people, is the kind of meta-expertise known as local discrimination. We know that local discrimination works and serves a vital function. This is where the journalists and whistle-blowers come into their own, not trying to do half-baked technical science, but fully cooked political investigation of science. Local discrimination is essential if science is to be kept honest and pure, and the anti-vaccination campaigners could play their part. But to be effective, they must do it properly and show that it has been done properly. Personal conviction is not expertise; and a good television personality is not expertise. Expertise in this regard is shown through hard and visible detective work and, as far as I know, it has not yet been done in the case of vaccination. The application of local discrimination is, I am suggesting, an important and legitimate way for the general public to contribute to scientific debate.

Default expertise

Now we come to the most difficult question. What if the emperors have no clothes? What if the bar is on the floor? And the bar certainly is on the floor in some sciences. It is the case that, armed with some idea of the possibilities, my guess or your guess about next year's inflation rate is as likely to be as right as that of the most highly paid and highly funded research econometricians with their complex computer models. Surely in such a case we can claim 'we are all scientific experts'. I am going to argue that it is not so, but here the argument gets hard and, therefore, especially interesting.

There is one obvious way in which the econometricians are experts and you and I are not; the econometricians have the computer models. They do not give the right answer but that does not mean the econometricians know nothing. They know the relationships between the variables in the models even if the complexity of the entire problem is beyond them. So, arguably, they are in a better position to make good judgements even if they are not based on the computer models but rather their experience of working with the materials. Furthermore, if we stick with the econometricians, they may improve and learn to do better over time whereas you and I will not. So those are two reasons for thinking that the econometricians might be worth something, even if the bar is on the ground in respect of their predictions. The third reason is more profound, more important, and returns us to our starting point, the zeitgeist – the way we think about scientists.

An anthropologist who studied the anti-vaccination campaigners and reported their impact on parents had something to say about the zeitgeist which we can probably all recognize: 'With the explosion of "contrary" expertise online, Kaufman says, "many parents see even the most respected vaccine expert's perspective on the issue as just one more opinion".'[46] But scientists are not people who have 'just one more opinion'. Before and just after the Second World War, under what we have called 'Wave 1' of science studies, it was thought that scientists had a special kind of opinion because their work led to the most true and efficacious results; science stood above the plain of other ways of being in the world when it came to accessing the workings of nature. Wave 2 of science studies levelled the cultural plain and eroded 'mount science'. Wave 2 reflected and, perhaps, fed into the changing zeitgeist. Now we need Wave 3. It is a straightforward social fact that not all groups and professions are the same. Consider: the Catholic Church has recently been beset by scandals concerning paedophile priests. The furore has been huge and so it should be because priests are meant to be special people who have sacrificed their family lives to care for their parishioners, so for a priest to abuse a child is seen as still worse than for an ordinary person to abuse a child, terrible though that is. But in spite of this scandal, if one was looking for a kind and good person, where would one look if not towards a religious leader? As always there are exceptions – individuals like the paedophile priests, certain entire religions and the overzealous activities of missionaries and those who preach terrorism – but, on the whole, one can say that the religious life is one of self-sacrifice and dedication to others. Compare religious leaders with bankers. Bankers have been taught that 'greed is good'; they have been taught that they are a very special kind of person for whom, unlike the rest of us, it is impossible to do a good day's work in the absence of astronomical bonuses; they really believe they are differently constituted from the rest of us in this regard and one can only imagine how frightened they must be when they come to be treated by, say, nurses who, they must believe, are incapable of doing a good job in virtue of their low pay and absence of vital end-of-year rewards. There will, once more, be exceptions: there will be good self-sacrificing bankers but, on the whole, the way of life of bankers is different from the way of life of religious leaders. If we were looking for a kind and good person we would not advise 'seek such a person among bankers'. We may ask this kind of question of all professional and other groups and most of us have enough ubiquitous discrimination to know how to answer it.

Now ask the question of scientists and we see that the answer many people will give is that scientists are just ordinary people with no special qualities; this is where the changing zeitgeist has taken us. But the answer is wrong. A sociological understanding of science shows it is wrong. In fact, scientists are a special group of people. Even if, as a result of Wave 2, we no longer think of them as special in virtue of their relationship with nature, they are still special in terms of the values that drive their lives and their aspirations in respect of how they live their lives. Though it is true that when scientific facts are judged it is a matter of judging scientists as much as their data, the scientists regret it; they are ashamed of it, not proud of it. Climategate shows this; it shows the scientists were ashamed because they had been seen to be judging scientists instead of facts, and the public outcry showed that the public expected more of scientists – they expected them to act as a special group of people. If Climategate had revealed the private emails of politicians, bankers or business leaders, there would have been no surprise because politicians, bankers and business leaders are expected to be judging people; they have a different kind of ethos. In contrast, when we discover that the tobacco companies' supposedly scientific results are actually driven by a financial or political motivation we are horrified. We are horrified because this is not how science is supposed to be. That is why there is a special role for investigative journalists and ordinary people in examining scientists. The role is different when it comes to investigating politicians and business leaders because we expect their choices to be driven by financial and political motivations. The journalists should have been examining Wakefield and his connections, something for which they did have appropriate skills, not his science, for which they manifestly did not. Science is, then, expected to be different from politics and business, and mostly it is.

To understand science one has to set aside scientific fraudsters, scientists who are driven primarily by greed and scientists who are driven by fame. We have to set aside the muscular capitalists who proclaim that science is primarily about the generation of wealth. We have to set aside the theoreticians who allow themselves limitless licence to speculate, and the wild-eyed Darwinists. We have to set aside media scientists, lobbying scientists and small groups who meet to slap each other on the back because they are all so clever and can see the future. Like the paedophile priests, none of these represent what the profession of science is about. We have to look for the communities that are truly representative of science. We have to look at theorists who are ready to work out what measurements will support or falsify their theories. We have to look for the careful observers and experimentalists who are driven by finding out what makes up the world. We have to look for those whose prime goal is to find out what the world is made of.

The field of gravitational-wave detection, in which I embedded myself, seems to me to exemplify the heartlands of professional science. Among that community there are one or two scientists ready to embroider their speculations for the sake of winning a research grant, but nearly all of them are concerned only with doing good work and finding the truth of the matter. Indeed, my most recent study of the group led me to 'tear my hair out' with frustration at their determination to go not a fraction further than their evidence. They had found something that should have been the culminating triumph of half a century of searching and they refused to announce it as a discovery because they were a fraction less than sure than they thought they ought to be.[47] Even for the burningly ambitious scientist, the aspiration is to discover something new; to cheat or bend the data is to sacrifice the thrill of discovery and the possibility of permanent recognition for a momentary, mundane, reward. Scientists may not have taken a vow of poverty but, more than riches, they want to be remembered in the way Einstein and Darwin are remembered, the scientific version of a heavenly afterlife. To cheat is to ensure that there can be no such lasting memory. Thus, integrity is built into the very nature of science and the relatively rare cheats have simply made a mistake and sacrificed their scientific birthright.

The Wave 1 sociologist Robert Merton mostly grasped this point. Merton set out the 'norms of science' that he thought made science work as it did. For example, he said that the community was 'universalistic': scientists' claims had to be judged without reference to their ethnicity or religious beliefs. He said that science depended on 'organized scepticism' whereby a scientist's work was subject to the critical scrutiny of his or her peers. He argued that science was characterized by 'disinterestedness', meaning that results were not to be determined by personal interest.[48] Merton claimed that if science did not abide by these norms it would not be so efficacious in uncovering the secrets of nature. In this last point, Merton was wrong. Wave 2 studies have shown that the norms are not always followed and that efficacious science can be carried out without cleaving to the norms. Nevertheless, so long as scientists aspire to follow the norms, the community can still be characterized by its special ethos. There are many more norms that can be listed but enough has been done to make the main points.[49] The main points are that the scientific community is characterized by its aspiration to live by a set of norms and, irrespective of their efficacy in producing results, the norms are good in themselves – good in a moral sense. To see this, consider just the three Mertonian points we have outlined plus the norms of honesty and integrity. Honesty and integrity are good in themselves. Universalism is good in itself. To expose one's ideas to the scrutiny of others is good in itself. And to act disinterestedly in the pursuit of knowledge is good in itself. Remember the target diagram: this is what characterizes the inside of science as opposed to the polarized campaigns which take over as we move out from the core and distance begins to lend enchantment.

Now compare this with anti-vaccination campaigns. The anti-vaccine campaigners are notorious for citing scientific material that favours their cause irrespective of where it is published: fringe journals and the Internet are counted equally with the mainstream while articles that take an opposite view or refute a favoured anti-vaccine claim are ignored, not on the basis of the reputation and past work of the scientist, but reading backwards from conclusions they reject. Of course they do such things: they are campaigners, not scientists!

Science, in contrast, is even a good institution when it makes its judgements about *scientists* as opposed to science. When Joe Weber came to be ignored by his peers it was not because he had written a letter to his Congresswoman demanding that the big project currently under way be stopped (which he did); it was because his peers thought he had long been wrong in his scientific claims. And when Joe Weber, in the 1980s, put forward new claims, concerning not gravitational waves but neutrino-detection, they were published and discussed in the mainstream literature (though eventually dismissed) in spite of his lack of credibility in a closely related field. Scientists judge other scientists badly because of their scientific judgements, not their interests or personal judgements – at least, that is how the ideal scientists I am talking about aspire to make their judgements. In contrast, an anti-vaccine campaigner such as Jenny McCarthy will one day pass her 'sell-by date' just as science does, but it will not be because what she espouses has lost scientific credibility; it will be because she is no longer publicity-worthy as rated by the general, television-viewing, public. Choose your community!

Now let us return to the world of the inefficacious econometric modeller where the bar is on the ground. It remains that the econometricians, so long as they are good ones, belong to the world of science and it follows that any judgements they make will be the judgements of good people. Of course, the journalists and other scrutineers, including their fellow economists, may show it is not so, but let us imagine this does not happen. The same kind of thing can be said of long-term weather forecasters and those who practise all the complex sciences that are unable to make accurate judgements; they are trying to find the truth of the matter and, all being well, trying to do it by cleaving to good values. As far as all these sciences are concerned, it is still the case that 'we are not all scientific experts now' because we do not belong to the scientific community and we do not necessarily make our judgements from the platform of the norms and aspirations that drive that community. We, ordinary mortals, are likely to be less honest, less universalistic, less ready to expose ourselves to expert criticism and less disinterested – unless we are serious whistle-blowers.

In sum

We may all be experts in this way or that but we are not all *scientific* experts now. We are all *ubiquitous experts* but this counts for nothing where serious scientific disputes are in question. We are all specialist experts in our jobs but this does not make us *scientific* experts except in special circumstances. If we start to believe we are all scientific experts, society will change: it will be those with the power to enforce their ideas or those with the most media appeal who will make our truths, according to whatever set of interests they are pursuing. The zeitgeist has to change if we want to preserve society as we know it because we have to start raising the value of plain ordinary science in our minds. Nevertheless, without changing society, small numbers of initially ordinary people can become scientific experts, not through reading but through specialist experiences at work, or experience of chronic diseases, or extended discourse with existing experts. There are, however, no short cuts and the Internet and even the professional journals are not reliable sources of scientific expertise. We are not all *meta-experts* when it comes to judging technical matters, though, once more, some of us through assiduous investigative work on local circumstances can become such and play a valuable role in safeguarding the norms of science. And we are not all *default experts* because we do not share the scientific ethos which may be the most valuable contribution of science to society. Science is inexact, messy and more often a matter of judgement than calculation. But if we want our judgements about the natural and social worlds made by good, disinterested people, then we should start the zeitgeist moving in the other direction and learn, once more, to elevate science to a special position in our society. Under this model, of course, the scientists too have a lot to live up to!

1 See Evans, 1999, for an analysis of the econometric modellers.
2 Bizarrely, Fish is reported to be saying nowadays that he was talking about Florida not the UK – odd, as he can be seen standing in front of a map of the UK in the course of his forecast (<http://www.youtube.com/watch?v=uqs1YXfdtGE> – accessed 21 February 2013).
3 Sociology of scientific knowledge (see chapter 1) shows that these triumphs were much less straightforward than they are presented as being, but let us leave that aside for now.
4 Michaels, 2008. The kind of understanding of how science works, developed in the 1970s by the subject known as science studies, specifically, Wave 2 of science studies (see chapter 1), might well have been responsible for showing the tobacco companies and the oil industry that it is easy to create a scientific controversy and how effective such a thing can be. It can show the interested parties that scientific argument is never decisive and that it is always possible to re-open what looks like a closed debate, and it provides some clues for how to do it.
5 Oreskes and Conway, 2010.
6 At: <http://republicans.globalwarming.sensenbrenner.house.gov/Press/PRArticle.aspx?NewsID=2740>.
7 Tart, C. T., 1972, 'States of Consciousness and State-Specific Sciences', *Science* 176: 1203–10. A paper by Paul Forman, 1971, argued that the zeitgeist of Weimar Germany, characterized by uncertainties as it was, naturally gave rise to quantum theory.

It is hard to say if the zeitgeist really can affect physics in this way but it is an intriguing possibility. It is certainly much easier for it to affect the thinking of social scientists and philosophers.

8　For an accessible account of the Michelson–Morley experiment and many of the other passages of science discussed in this book, see Collins and Pinch, 1993/1998.

9　Bloor, 1973. Bloor drew explicitly on Wittgenstein and, we must assume, implicitly on the atmosphere of the times.

10　Collins, 1974, 1975, 1985/1992.

11　The search for gravitational waves is still going on with apparatus many millions of times more sensitive, but which still exhibits the kind of uncertainties surrounding the early detectors (see, for example, my 2013a book, *Gravity's Ghost and Big Dog*).

12　What this work does not show is that scientific findings are simply the result of fashion. It shows that in the short term it is hard to produce a clear outcome to a deep scientific controversy using scientific methods – theory and experiment – alone. We do not know what drives scientific beliefs in the long term. What we can say is that there is less science and more 'politics' than previously thought at the beginning of scientific controversies and that we should not be surprised that many such controversies continue for half a century or more, at least in underground form.

13　The chief and very successful driver of this trend was Bruno Latour.

14　It might be said that this is to confuse science and technology – some scientists love to say this so as to keep their science pure. But, first, experiments are mostly technology so the distinction may be viable in theory but not in practice. Second, as far as the public is concerned, there is no distinction: one's fridge is as much a product of science as one's black hole. There are a group of sociologists of technology, however, who point out that technology is different in that the public have a much greater role in determining what is a successful technology and what is not.

[15] Irwin, 1995.

[16] Wynne, 1996.

[17] Some sense of where the tendency led can be gained from the fact that researchers of the 'lay expert' persuasion continue to defend the ordinary mothers and fathers who, in the UK, revolted against MMR vaccine in response to the completely unfounded claim that it could cause autism. This argument still goes on in science and technology studies.

[18] Epstein, 1996.

[19] Turner, 2003. See also Turner, 2001: 'The phenomenon of expertise produces two problems for liberal democratic theory: the first is whether it creates inequalities that undermine citizen rule or make it a sham; the second is whether the state can preserve its neutrality in liberal "government by discussion" while subsidizing, depending on, and giving special status to, the opinions of experts and scientists.'

[20] Jasanoff, 2003, pp. 397–8. This and the following are, of course, only isolated quotations from a valuable body of work on how different democracies handle the relationship between experts and the public. Nevertheless, they do illustrate the points perfectly.

[21] Jasanoff, 2003, p. 397.

[22] Huxham and Sumner, 1999.

[23] At: <http://en.wikipedia.org/wiki/John_von_Neumann#Cognitive _abilities> (accessed 23 February 2013).

[24] Beller, 1999, pp. 259–61.

[25] Of course, one would still not refer to these contributors to science as 'scientists' in virtue of their observations. See also Collins, 2013b.

[26] Simplified and adapted from Collins and Evans, 2007, p. 14.

[27] First being discussed in Collins and Evans, 2002.

[28] Giles, 2006.

[29] Collins's answer is on the right and was quickly worked out by thinking about the geometry of waves and detectors.

[30] See Collins and Sanders, 2007.

[31] 'Meta-criteria' are given their own row in the original version of the table in Collins and Evans, 2007.

[32] For the notion of 'core-set' and a more detailed working out of this process, see Collins, 1985/1992.

[33] Donald Mackenzie, who took over this idea (e.g., 1997), pointed out that funders and policymakers are more aware of the uncertainties than the other outer rings and so there is a 'trough of certainty'.

[34] At: <http://www.cgd.ucar.edu/cas/Trenberth/statement.html> (accessed 28 February 2013).

[35] At: <http://www.skepticalscience.com/Mikes-Nature-trick-hide-the-decline.htm> (accessed 28 February 2013).

[36] We will see that there are other kinds of grounds that might be justified on special occasions.

[37] Indeed, in spite of the fact that several generations of new and much more sensitive detectors have been built, the consensus is that the waves have still not been seen.

[38] At: <http://news.bbc.co.uk/1/hi/8388485.stm> (accessed 12 March 2013).

[39] Much of my understanding of the Mbeki case is drawn from discussions and the written works of Martin Weinel; see Collins and Weinel, 2011; Weinel, 2008, 2009, 2010, 2012. And see Nattrass's important 2007.

[40] Weinel, 2010; Michaels, 2008; Oreskes and Conway, 2010.

[41] In the spring of 2013, in Swansea and West Wales, 40 miles west of where this book is being written, there is a fast-growing measles epidemic.

[42] See: <http://en.wikipedia.org/wiki/MMR_vaccine_controversy>; <http://news.bbc.co.uk/1/hi/health/1808956.stm> (accessed 12

March 2013). Though I am citing the Internet for readers to access, my account is based on my own experience of living through the events.

[43] Gross, 2009.

[44] Davies et al., 2002, found nearly all [anti-vaccination] sites adopted an 'us versus them' approach, casting doctors and scientists as either 'willing conspirators cashing in on the vaccine "fraud" or pawns of a shadowy vaccine combine'. Parents' intuitive views about vaccines were elevated above 'cold, analytical science' (quoted in L. Gross, 2009).

[45] '"Show Your Working": What "ClimateGate" Means' (<http://news.bbc.co.uk/1/hi/8388485.stm>).

[46] Gross, 2009, reports the work of anthropologist Sharon Kaufman, who studied the anti-vaccination campaigners.

[47] Collins, 2013a.

[48] And hence the scandal when it was discovered that Wakefield had a financial interest in the truth of his claims.

[49] The argument made here is being worked out in greater detail by the author and his colleagues under the heading of 'Elective Modernism'. Elective Modernism argues that we should choose science because it is morally good.

Beller, Mara. 1999. *Quantum Dialogue: The Making of a Revolution*. Chicago, IL: University of Chicago Press.

Bloor, David. 1973. 'Wittgenstein and Mannheim on the Sociology of Mathematics', *Studies in the History and Philosophy of Science* 4: 173–91.

Collins, Harry. 1974. 'The TEA Set: Tacit Knowledge and Scientific Networks', *Science Studies* 4: 165–86.

Collins, Harry. 1975. 'The Seven Sexes: A Study in the Sociology of a Phenomenon, or The Replication of Experiments in Physics', *Sociology* 9/2: 205–24.

Collins, Harry. 1985/1992. *Changing Order: Replication and Induction in Scientific Practice*, 2nd edn. Chicago, IL: University of Chicago Press.

Collins, Harry. 2013a. *Gravity's Ghost and Big Dog: Scientific Discovery and Social Analysis in the Twenty-First Century*. Chicago, IL: University of Chicago Press.

Collins, Harry. 2013b. 'Three Dimensions of Expertise', *Phenomenology and the Cognitive Sciences* 12/2: 253–273, DOI: 10.1007/s11097-011-9203-5.

Collins, Harry, and Evans, Robert. 2002. 'The Third Wave of Science Studies: Studies of Expertise and Experience', *Social Studies of Science* 32/2: 235–96.

Collins, Harry, and Evans, Robert. 2007. *Rethinking Expertise*. Chicago, IL: University of Chicago Press.

Bibliography

Collins, Harry, and Pinch, Trevor. 1993/1998. *The Golem: What You Should Know About Science*. Cambridge and New York: Cambridge University Press [2nd edn, Canto, 1998].

Collins, Harry, and Sanders, Gary. 2007. 'They Give You the Keys and Say "Drive It": Managers, Referred Expertise, and Other Expertises', in Collins (ed.), *Case Studies of Expertise and Experience: Special Issue of Studies in History and Philosophy of Science* 38/4: 621–41.

Collins, Harry, and Weinel, Martin. 2011. 'Transmuted Expertise: How Technical Non-Experts Can Assess Experts and Expertise', *Argumentation* 25/3: 401–13.

Davies P., Chapman, S., and Leask, J. 2002. 'Antivaccination Activists on the World Wide Web', *Archives of Disease in Childhood* 87: 22–5.

Epstein, Steven. 1996. *Impure Science: AIDS, Activism and the Politics of Knowledge*. Berkeley, Los Angeles, CA, and London: University of California Press.

Evans, Robert. 1999. *Macroeconomic Forecasting: A Sociological Appraisal*. London: Routledge.

Fleck, Ludwik. 1979. *Genesis and Development of a Scientific Fact*. Chicago, IL: University of Chicago Press [first published in German in 1935].

Forman, Paul. 1971. 'Weimar Culture, Causality and Quantum Theory, 1918–1927: Adaptation by German Physicists and Mathematicians to a Hostile Intellectual Environment', *Historical Studies in the Physical Sciences* 3: 1–115.

Giles, J. 2006. 'Sociologist Fools Physics Judges', *Nature* 442: 8.

Gross, L. 2009. 'A Broken Trust: Lessons from the Vaccine–Autism Wars', *PLoS Biol* 7/5: e1000114. doi:10.1371/journal.pbio.1000114 (p5).

Hargreaves, I., Lewis, J., and Speers, T. 2003. 'Towards a Better Map: Science, the Public and the Media', *Economic and Social Research*

Bibliography

Council. Available at: <http://www.comminit.com/en/node/177710>.

Huxham, M., and Sumner, D. 1999. 'Emotion, Science and Rationality: The Case of the "Brent Spar"', *Environmental Values* 8: 349–68.

Irwin, Alan. 1995. *Citizen Science: A Study of People, Expertise, and Sustainable Development.* London: Routledge.

Jasanoff, Sheila. 2003. 'Breaking the Waves in Science Studies; Comment on H. M. Collins and Robert Evans, "The Third Wave of Science Studies"', *Social Studies of Science* 33/3: 389–400.

Kuhn, Thomas S. 1962. *The Structure of Scientific Revolutions.* Chicago, IL: University of Chicago Press.

MacKenzie, D. 1997. 'The Certainty Trough', in R. F. W. Williams and J. Fleck, *Exploring Expertise: Issues and Perspectives.* Basingstoke: Macmillan, pp. 325–9.

Merton, R. K. 1942. 'Science and Technology in a Democratic Order', *Journal of Legal and Political Sociology* 1: 115–26.

Michaels, David. 2008. *Doubt is Their Product: How Industry's Assault on Science Threatens Your Health.* New York: Oxford University Press.

Nattrass, Natalie. 2007. *Mortal Combat: AIDS Denialism and the Struggle for Antiretrovirals in South Africa.* Scottsville: University of KwaZulu Press.

Office of Science and Technology and the Wellcome Trust. 2000. *Science and the Public: A Review of Science Communication and Public Attitudes to Science in Britain.* London: Wellcome Trust.

Oreskes, Naomi, and Conway, Erik, M. 2010. *Merchants of Doubt: How a Handful of Scientists Obscured the Truth on Issues from Tobacco Smoke to Global Warming.* London: Bloomsbury Press.

Snow, C. P. 1959. *The Two Cultures and the Scientific Revolution.* Cambridge: Cambridge University Press.

Bibliography

Tart, C. T. (1972). 'States of Consciousness and State-Specific Sciences', *Science* 176: 1203–10.

Turner, Stephen. 2001. 'What is the Problem with Experts?', *Social Studies of Science* 31: 123–49.

Turner, Stephen. 2003. *Liberal Democracy 3.0: Civil Society in an Age of Experts*. Newbury Park: Sage, p. 5.

Weinel, M. 2008. 'Counterfeit Scientific Controversies in Science Policy Contexts. Cardiff School of Social Sciences Working Paper #120'. Cardiff: Cardiff School of Social Sciences.

Weinel, M. 2009. 'Thabo Mbeki, HIV/AIDS and Bogus Scientific Controversies'. Politicsweb 19 March. Retrieved from: <http://www.politicsweb.co.za/politicsweb/view/politicsweb/en/page71619?oid=121968&sn=Detail>.

Weinel, M. 2010. 'Technological Decision-Making under Scientific Uncertainty: Preventing Mother-to-Child Transmission of HIV in South Africa' (PhD Thesis). Cardiff: Cardiff School of Social Sciences.

Weinel, M. 2012. 'Expertise and Inauthentic Scientific Controversies: What You Need to Know to Judge the Authenticity of Policy-Relevant Scientific Controversies', in J. Goodwin (ed.), *Between Scientists & Citizens: Proceedings of a Conference at Iowa State University, June 1–2, 2012*. Ames, IA: Great Plains Society for the Study of Argumentation, pp. 427–40.

Winch, Peter G. 1958. *The Idea of a Social Science*. London: Routledge and Kegan Paul.

Wittgenstein, Ludwig. 1953. *Philosophical Investigations*. Oxford: Blackwell.

Wynne, Brian. 1996. 'May the Sheep Safely Graze? A Reflexive View of the Expert–Lay Knowledge Divide', in S. Lash, B. Szerszynski and B. Wynne (eds), *Risk, Environment & Modernity: Towards a New Ecology*. London: Sage, pp. 27–83.

1960s 18, 22
245T 39

academics 18, 43, 45
AIDS and anti- retroviral drugs
    activists 41, 42
    in South Africa 75, 98–101
anti-retroviral drugs *see* AIDS
arXiv 34

Bohr, Niels 56
Brent Spa 47
BSE *see* 'Mad cow disease'

Calder Hall 1, 3
capital punishment 119
chronic diseases 64
circumcision 7
Clausewitz 38
Climategate 11–14, 77, 80–97,
    104, 125
core-set 83–7
'cranks' 36
crown jewels of science 10
Cumbrian sheep farmers 40

democracy
    paradox of 46
    sentiment for 41
    simple conception of 43–4

direct square law for knowledge 85
'distance lends enchantment' 25,
    85, 129
Duesberg, Peter 99

econometric modelling 7, 121f.,
    130
Einstein, Albert 23, 55
environmental movement 46
experimenter's regress 32, 39
expertise and experts
    10,000 hours model of 51, 58
    beer-mat knowledge 66
    contributory 65, 72
    default 15, 19, 41, 54, 59, 104,
        121
    downward discrimination 78,
        107, 110
    experience-based 39, 41
    interactional 68–73, 95–7, 116
    interactional expertise and
        primary source knowledge,
        gulf between 96
    'lay' 41, 64
    local discrimination 76–7, 103
    meta- 59, 74ff., 110, 119
    popular understanding 67
    primary source knowledge 67,
        95–6, 118
    referred 77

Index

relational theory of 45, 49, 60
specialist 58, 64, 116
substantive 50
technical connoisseurship 78
theory of acquisition of 59–61
transmuted 77
ubiquitous 53–6, 59, 63, 115
ubiquitous discrimination 75,
    110, 119–20, 124

feminist analysis 39
Fleck, Ludwik 23
foot-and-mouth epidemic 5

genetically modified organisms
    (GMOs) 47, 78
gravitational waves, detection of
    31–2, 69–70, 83, 93, 126

Hulme, Mike 97, 113, 119
hurricane, unexpected in UK 8

interpretative flexibility 36, 39,
    43, 47

Kaufman, Sharon 122
Kuhn, Thomas 23–6, 29

McCarthy, Jenny 111–12, 130
'mad cow disease' or BSE 4
managers, of large scientific
    projects 72
Marshall, Barry 92
Mbeki, Thabo 98–100, 104, 118
Merton, Robert 127–8
Michelson-Morley experiment 23
Midnight Cowboy 75
Mullis, Kary 99

necrotizing fasciitis 6
newspapers and journalists 108,
    116, 119
'Nobelskigrad' 56–8, 63
Nuovo Cimento 93, 98

oil companies 36

paradigms 33
paranormal phenomena, research
    on 33
peer review 72
placebo effect 6
Progress in Physics 34
    'declaration of academic
        freedom' in 35
Ravetz, Jerry 97, 113, 119
replication 29

science
    as a collective activity 91
    as a ubiquitous expertise 56;
        see also 'Nobelskigrad'
    counterfeit controversies in 100,
        104, 108
    heroes in 92
    mavericks in 92–3
    norms and values of 82, 87,
        124–8, 131
    war 88
science studies 20
    Wave 1 21–4, 36, 82, 122, 127
    Wave 2 24, 28, 36, 80–3, 123
    Wave 3 81, 123
scientific revolutions 26–8
semiotics 38
'show your working' 97–9, 113
sociology of error 29

sociology of scientific knowledge
  (SSK) 27–8, 37
SSK *see* sociology of scientific
  knowledge
Strauss, Lewis 3
Structure of Scientific Revolutions
  22
symmetry, principle of 28

tacit knowledge 31, 60, 69, 91
  ubiquitous 66
'talking the talk' 73
target diagram 83–5
TEA-laser 30–2, 60
television, science on 15
The Two Cultures 37
tobacco companies 10, 36, 103,
  110, 125
tonsillectomy 6–7

underdogs, preference for among
  sociologists 47

vaccination
  and herd immunity 111
  anti-vaccination campaigners
    104–5, 111–12, 118, 122,
    129
  autism and 105–6, 109, 113
  measles and 106
  mumps, measles and rubella
    (MMR) 105, 106, 108–10,
    116
von Neumann, John 55

Wakefield, Andrew 105–8
Warren, Robin 92
Weber, Joe 93–5, 129
whistle-blowers 103, 120, 131
Winch, Peter 25
Wittgenstein, Ludwig 25
word processors 52

zeitgeist 18, 43, 45, 122, 124
ZETA 1, 3, 4